Java Progi

Java Programming for Spatial Sciences

Dr Jo Wood
City University
London

London and New York

First published 2002
by Taylor & Francis
11 New Fetter Lane, London EC4P 4EE

Simultaneously published in the USA and Canada
by Taylor & Francis Inc,
29 West 35th Street, New York, NY 10001

Taylor & Francis is an imprint of the Taylor & Francis Group

© 2002 Dr Jo Wood

Publisher's Note
This book has been prepared from camera-ready copy supplied
by the author
Printed and bound in Great Britain by
The Cromwell Press, Trowbridge, Wiltshire

All rights reserved. No part of this book may be reprinted or
reproduced or utilised in any form or by any electronic,
mechanical, or other means, now known or hereafter
invented, including photocopying and recording, or in any
information storage or retrieval system, without permission in
writing from the publishers.

Every effort has been made to ensure that the advice and information
in this book is true and accurate at the time of going to press.
However, neither the publisher nor the authors can accept any legal
responsibility or liability for any errors or omissions that may be made.
In the case of drug administration, any medical procedure or the use of
technical equipment mentioned within this book, you are strongly
advised to consult the manufacturer's guidelines.

British Library Cataloguing in Publication Data
A catalogue record for this book is available
from the British Library

Library of Congress Cataloging in Publication Data
A catalog record for this book has been requested

ISBN 0–415–26098–1 (pbk)
ISBN 0–415–26097–3 (hbk)

Contents

List of Figures   viii
List of Tables   x
Preface   xi
Acknowledgements   xiii

1 Introduction   1

1.1 Welcome 1
1.2 What is a Programming Language? 4
1.3 Programming Styles 5
1.4 Object-Oriented Modelling 7
1.5 Why Program With Java? 10
1.6 A Short History of Java and the Internet 11
1.7 Creating a Working Program 13
1.8 Examples 15
1.9 Summary 18

2 Introducing Classes and Objects   19

2.1 Class Design 19
2.2 Classes and Objects 22
2.3 The State of a Class 23
2.4 The Behaviour of a Class 31
2.5 Making Code Clear 33
2.6 Distinguishing Classes from Objects in Java 35
2.7 Summary 40

3 Developing Classes and Objects   43

3.1 Inheritance 43
3.2 Abstract Methods and Interfaces 48
3.3 Passing messages 50

3.4 Using Graphical Classes in Java 55
3.5 Case Study: Modelling Ants In The Garden 62
3.6 Summary 72

4 Controlling Program Movement   73

4.1 Procedural and Object-Oriented Design Revisited 73
4.2 Applying Operators to Variables 73
4.3 Rules of Precedence and Precision 77
4.4 Controlling Movement With loops 78
4.5 Grouping Data in Arrays 84
4.6 Laying Out Graphical Components 89
4.7 Case Study: Creating a Displayable Raster Map 93
4.8 Summary 101

5 Making Decisions   103

5.1 Making Decisions With `if` 103
5.2 Making Decisions With `switch` 111
5.3 Other ways of Making Decisions 115
5.4 Good Decision Making Design 116
5.5 Case Study: Adding Spatial Classes to Garden Ants 117
5.6 Summary 129

6 Sharing Classes   131

6.1 Controlling Variable Scope 131
6.2 Documenting Java code 135
6.3 Java Packages 137
6.4 Case Study: Creating A Graphical Ants Model 142
6.5 Summary 153

7 Collecting Objects Together   155

7.1 Vector Modelling of Spatial Objects 155
7.2 Dynamic Groups 159
7.3 Case Study: Feeding Ant Colonies 173
7.4 Summary 190

8 Controlling Dynamic Events   193

8.1 Event Handling 193
8.2 Threads 200

    8.3 Case Study: Creating a Displayable Spatial Model 210
    8.4 Summary 230

9 Handling Streams and Files   233

    9.1 Input and Output Streams 233
    9.2 Reading and Writing Files 240
    9.3 Analysing Strings 247
    9.4 Object Serialization 249
    9.5 Case Study: Adding File Handling to Spatial Objects 251
    9.6 Summary 263

10 Communicating with the Wider World   265

    10.1 The Applet 265
    10.2 Case Study: A Simple Map Applet 274
    10.3 Communicating Using XML 279
    10.4 Case Study: Converting Applications into Applets 289
    10.5 Summary 299

References and Further Reading	301
Glossary	305
Index	315

Figures

1.1	The high-level – low-level programming continuum	4
1.2	A Turing machine	5
1.3	A simple ants' nest model containing only one ant	9
1.4	A More complex model with many ants	9
1.5	Compiling and running a Java program on Windows and Unix platforms	15
1.6	Output from `Hello2.java`	18
2.1	A map of a simple Park	20
2.2	Class diagram representing the park	20
2.3	Class diagram of the park model	36
3.1	Class diagram for Park with inheritance	44
3.2	Passing messages to and from an object	51
3.3	The structure of a mutator method	53
3.4	The structure of an accessor method	53
3.5	Sample AWT components	55
3.6	The Abstract Windowing Toolkit (AWT) component hierarchy (simplified)	57
3.7	The Swing component hierarchy (simplified)	59
3.8	Swing components inherited directly from `JComponent`	60
3.9	Output from `SimpleWindow.java`	61
3.10	Class diagram showing simple design for ants in the garden model	65
3.11	Ants in the garden class design ready for coding in Java	67
4.1	A raster image showing the arrangement of raster cells	87
4.2	A simple raster model of space	87
4.3	A window with three components controlled by the layout manager	91
4.4	Five buttons arranged using a `FlowLayout` manager	91
4.5	Five buttons arranged using a `BorderLayout` manager	92
4.6	Five buttons arranged using a `GridLayout` manager	92
4.7	Sketched design of a raster display GUI	94
4.8	A 2D raster to 1D image transformation	97
4.9	Hierarchical arrangement of graphical components in the `RasterDisplay` GUI	99

Figures

4.10	The final raster display window	101
5.1	A 2D spatial footprint	118
5.2	Garden ants class diagram incorporating the `SpatialModel` interface	123
5.3	Six topologic relationships between two rectangles	124
6.1	Part of the javadoc output for the Park classes	136
6.2	Three reusable packages inside the `jwo.jpss` package	141
6.3	Class diagram showing drawable ant simulation	152
6.4	Graphical output from the simple ant simulation	152
7.1	Vector representation of simple linear and areal features	156
7.2	Two vector objects and their bounding rectangles	158
7.3	Some Java collections	161
7.4	A collection of `GISVectors` assembled in a `VectorMap`	166
7.5	A simple Gazetteer table using two classes to store the keys and values	169
7.6	Ant class diagram with queen and breeding behaviour	186
7.7	Six stages in the ant simulation	190
8.1	A simple window that responds to mouse click action events	195
8.2	Simple Text editor window with menu that responds to mouse selection	198
8.3	The sequence of two-threaded processes in the Quiz class	206
8.4	Output from the Prime Number Explorer	210
8.5	Classes and interfaces in the `jwo.jpss.spatial` package	211
8.6	Improved `jwo.jpss.spatial` package containing GUI classes	215
8.7	Sketched Design for Spatial Model Display	216
8.8	Pixel and georeferenced coordinate systems	219
8.9	Some Java2D rendering styles and strokes	224
8.10	Point-in-polygon testing using `GeneralPath`'s `compare ( )` method	226
8.11	Examples of `ModelDisplay` output using a `RasterMap` and `VectorMap`	230
9.1	Example of a File Chooser used to select a file to open	245
9.2	Example of error message reported if selected file is not found	245
9.3	File chooser with an image filter	247
9.4	Spatial class diagram with serialisation	257
9.5	Sample output from `ModelDisplay` showing file handling	263
10.1	Example of a simple text editor applet	266
10.2	Map interrogator Applet	279
10.3	A DOM tree structure of a simple XML file	282
10.4	Sample output from ant simulation applet status at bottom of window	294
10.5	Output from the Spatial Model Display Applet	299

Tables

2.1	Examples of abbreviated variable names	26
2.2	Java primitives	27
4.1	Abbreviated arithmetic operators	75
4.2	Logical operators	77
6.1	Four types of Java packages	138
6.2	Some commonly used Java packages	139
6.3	Drawing procedure for the ant simulation	143
7.1	Methods in the `Collection` interface	162
7.2	Methods in the `Iterator` interface	163
7.3	Ant genes	182
8.1	Methods in the `WindowListener` interface	200
8.2	Spatial classes and interfaces previously developed	210
9.1	The *ArcGrid* ASCII file format	252
9.2	Generic (*GRASS*) vector map file format	252
10.1	Some classes to avoid when creating Java 1.1 applets	273

Preface

In 1992, as a new lecturer in Geographic Information Systems at Leicester University, I was asked to teach a postgraduate course in programming for GIS. Two years previously I had taught myself the programming language C as part of my masters thesis, so I thought it might be a 'Good Idea' to develop a 'C Programming for GIS' module. That Good Idea soon turned into far too many hours in front of a C compiler and word processor, trying to distil the essence of a complex low-level programming language in terms that would be relevant to non-programming GIS students. Despite the plethora of C textbooks at the time, there was none with an emphasis on the kinds of programming problems faced by those with spatial applications in mind. I was forced to provide detailed web-based notes illustrated with plenty of programming examples that could be downloaded by students.

As the years passed, those notes became increasingly detailed and C evolved into C++ and eventually Java. Yet still one of the most frequently asked questions by GIS students a the beginning of a programming course was 'What is the best book for programming C/C++/Java?'. There were of course, plenty of introductory programming texts, and over time, plenty of books covering Geographic Information Science, but no texts that bridged this apparently unbridgeable gap. It was probably in an idle moment soon after learning the Java language in 1996 that 'Good Idea Number 2' emerged and has been nagging me ever since. Why not take this new language, which many people seem quite excited about, and adapt those lecture notes to make a book? As part of the annual process of updating my lecture notes, I started to write them as if they were pages in a text book with the idea that if a publisher ever express an interest in the idea, I would simply 'cut and paste' a book within a matter of weeks.

It is of course, exactly this kind of naïve optimism that gets an awful lot of us in trouble, yet at the same time, is the only way we can delude ourelves that we are capable of achieving such tasks. I would almost be tempted to say that writing a book is like having a baby – if we really knew how much effort, pain and suffering was involved, we would never try in the first place. However, having very little experience of one, and no experience of the other, I will refrain from such glib comparisons.

So, as I sit here looking at the final draft of this book, realising that 'a few weeks' is a long time in publishing years, with my eyesight and posture considerably worse for wear, I can only hope that 'Good Idea Number 3' will be a long time coming.

Acknowledgements

In putting together a book based on work of the last 10 years, I owe my debts to many people. I would like to thank David Unwin, Alan Strachan and Peter Fisher who all, while I was a lecturer at Leicester, provided support and encouragement to pursue my programming efforts. Pete in particular gave me the confidence (and contacts) to pursue publishers with Good Idea Number 2. The hundreds of students both a Leicester and City Universities who have been forced (and sometimes even chosen) to take my programming courses have provided invaluable experience, encouragement and feedback. Certainly without them, this book would not have been written.

Thanks also to Tony Moore at Taylor and Francis who took the idea of a Java book for spatial sciences seriously, providing a great deal of encouragement at an important time. This book would not have been completed (this decade) were it not for Sarah Kramer at Taylor and Francis who has perfected the art of knowing when to push authors and when to leave them alone.

More recently, I owe a huge debt of thanks to colleagues and students at City University to whom I should have been paying more attention instead of working on what has become known as the 'B thing'. I would particularly like to thank Jason Dykes, Jonathan Raper and Susannah Quinsee who have all provided support, understanding and space when I most needed it.

I must also thank Jacqui Russell who has provided thought provoking discussion on Turing machines and ant behaviour, but above all a thoughtful, supportive and understanding environment for me to complete this book.

Oh, and thank you to the countless ants that have been busy going about their lives in my garden in London. You have provided a rich source of metaphor, distraction and object-oriented behaviour.

Jo Wood, October, 2001.

To Frank Burke
who taught me how to think properly

CHAPTER ONE

Introduction

1.1 WELCOME

This book is written for those who have an interest in getting computers to perform tasks that solve particular problems. Of course, most computer use is intended to perform such tasks, whether sending e-mail or using a Geographical Information System (GIS) to print a paper map. What we hope to develop in the pages of this book is the ability to express a problem in a way that a computer understands, allowing us more precision and control over what computers do for us.

The Java programming language provides us with a mechanism for both organising our thoughts and communicating with computers. So, the general aims of this book are twofold:

1 *To develop a way of thinking* in order to be able to take a problem and express its solution using object-oriented modelling.
2 *To build-up knowledge of the Java language* in order to be able to solve the types of problems faced when dealing with spatial information.

1.1.1 The Intended Audience

This book is written for novice programmers as well as those with some experience of programming in other languages. It has arisen from the author's 10-year experience of teaching programming to geographers and information scientists, most of whom had no previous programming experience. While no experience of handling spatial information is required, it is assumed that readers have some motivation for doing so. So, in no particular order, here are six readers who might benefit from reading this book.

Sandra is a postgraduate student studying for a degree in Geographic Information Science (GI Science). She has used a GIS for a few months but wishes to have a more detailed and critical understanding of how the computer software handles GIS data.

Orlando is an archaeologist who has spent the last 3 years collecting data on artefact locations in Southern Spain. He has never programmed before but would like to place his data 'on the web' and allow others to perform spatial analysis on them.

Sheila is a geomorphologist who learnt to program in Fortran 20 years ago and wishes to update her programming skills, which she hopes to be able to apply to her ongoing research on landscape change.

Jay is a social scientist who is used to using statistics packages to analyse her data on social exclusion. She suspects that there are spatial patterns to her data but she finds her statistics software rather limited in terms of spatial mapping and analysis.

Leo works for a company providing bespoke paper maps for clients. He wishes to develop a simple software application that allows clients to browse and purchase digital copies of his maps.

Brian is interested in mobile communication technologies and wishes to explore how they might be used to provide location-based services to handheld devices. He would like to know more about how Java can be used in this context.

1.1.2 How to Use this Book

If you have never programmed in any language before, the best way of using this book is to start at the beginning and work through the chapters in sequence. Each chapter builds upon the skills and knowledge of the previous ones. For those who wish to concentrate on particular aspects of Java programming, there are also a number of 'threads' which run through the book.

Graphical techniques and user interface design. Sections 3.4, 4.6, 4,7, 6.4, 8.3, 10.2, 10.4
Spatial Data Modelling. Sections 4.5, 5.5, 7.1, 7.2, 8.3, 9.5, 10.4
Ants in the Garden – Object orientation case study. Sections 1.4, 3.5, 5.5, 6.4, 7.3, 10.4
Java API – Useful Java packages, Sections 3.4, 7.2, 8.3, 10.3

This is not a book that can be read through in an hour in the bath, as it is as much about 'doing' as it is reading. The accompanying website at www.soi.city.ac.uk/jpss includes exercises and quiz questions as well as further source code examples to download. By interacting with the materials both in the book and on the web, you will develop your programming skills and 'way of thinking', both vital to the successful manipulation of computing technology.

Those who have some programming experience may feel tempted to skip the early parts of the book. Of course, you are free to do so, but you may find it useful to attempt the exercises associated with each chapter, just to make sure you have not missed out on any key concepts.

To encourage an interactive use of this book, you will find short questions and simple exercises throughout its pages and on the web. You are encouraged to attempt these before moving on as they are designed to help reinforce key ideas.

> *Short questions will appear in a box like this. For example, consider why you wish to learn the Java language. Which of the six readers identified above (Sandra, Orlando, Sheila, Jay, Leo and Brian) do you most closely identify with?*

At various stages, you will also come across 'asides' that explain related topics. These can be safely skipped without seriously damaging your health but are provided where it is thought that the background information might be useful in understanding the programming context.

> **Asides**
>
> Asides look like this. The Chambers English dictionary defines an aside as 'words spoken in an undertone, so as not to be heard by some person present, words spoken by an actor which the other persons on the stage are supposed not to hear: an indirect effort of any kind'.

To encourage further interactivity, you can find a website to support this book at

`http://www.soi.city.ac.uk/jpss`

Here you will find all the source code examples used in the book along with quiz questions, exercises and further Java resources to browse and download.

Frequent use of examples of Java code is made throughout this book. All Java code is shown in the Courier font in order to distinguish it from the commentary. Extended examples are shown as below and can also be downloaded from the website. Increasing reliance on examples will be made as you progress through this book and develop your Java reading skills.

```
//    *****************************************************
/** Class for representing trees.
  * @author    Jo Wood
  * @version   1.0, 14th May, 2001
  */
//    *****************************************************

public class Tree
{
    // -------------- Class and object variables -----------------

    private float height;   // Stores height of the tree.


    // -------------------- Constructor ------------------------

    /** Creates a new tree and gives it an initial height.
      */
```

```
    public Tree()
    {
        height = 1;          // Set initial height at 1m.
    }
}
```

Finally, a glossary of terms can be found at the back of the book. This can be particularly useful if when reading the book you come across a term you have forgotten and wish to remind yourself of its meaning.

1.2 WHAT IS A PROGRAMMING LANGUAGE?

A natural language such as English, Portuguese or Hindi can be regarded as a combination of a vocabulary and set of grammatical rules that together facilitate communication. In order to be able to communicate, we need some idea of both the vocabulary and grammar. Language allows us to assemble our own ideas, convey them to other people and receive them from others. It provides a common framework for the organisation of ideas that two or more people can share.

A programming language is similar in that it allows communication between ourselves and the computers that share the common language. Even without communicating with a computer, programming languages can be useful to us as they allow us to assemble ideas in a structured way that can make future communication with a computer easier.

At the most detailed internal level, computers can only understand one language – binary instructions consisting of 1s and 0s (known as *machine code*). Most of us, on the other hand, cannot. We tend to think in far more abstract and ambiguous terms. Therefore, we need a 'language' that both the computer and ourselves can understand. We also need something to translate this language into the binary numbers that the computer works with.

Programming languages can therefore be placed on a scale somewhere between the computer's representation of information and our own (see Figure 1.1).

Figure 1.1 The high-level – low-level programming continuum.

Introduction

High-level languages can be regarded as ones that are closer to our way of thinking and organising ideas. They have the advantage of being relatively easy to use as they require less effort on our part to move between our own conceptualisations and those forced on us by the high-level language. They have the disadvantage of often being less flexible as they are usually designed to be applicable to a particular set of tasks (for example, a spreadsheet is useful for manipulating numerical information, but less so for the display of large amounts of text). You can think of using a GIS as programming in a high-level language because many of our ideas about spatial representation and analysis are already 'encoded' in the language of the GIS.

Low-level languages are closer to the binary representations used by all computers and are therefore more easily and efficiently translated into machine code. They tend to be more difficult for us to understand, but they can be flexible and powerful.

Java sits somewhere in the middle of the continuum containing many high-level concepts (such as objects, sounds and images) but structured in such a way as to allow efficient translation into a lower-level language.

1.3 PROGRAMMING STYLES

In the 1930s, the mathematician Alan Turing conceived of a machine, a machine that he (at that time) had no intention of building, nor indeed did he suggest, could be built (Figure 1.2). This machine, he imagined, consisted of a limitless stream of boxes arranged in a line, each of which was capable of storing a symbol of some kind. Associated with this line of boxes was a device that was able to identify the symbol in any single box, and, if necessary, substitute it for a new symbol. This device was capable of moving from one box to an adjacent one either to the 'left' or 'right'.

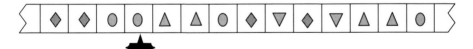

Figure 1.2 A Turing machine.

The behaviour of the moveable device was governed by a series of *rules* that determined its movement and whether it changed the symbol stored in the box it was scanning. So, for example, we might define the following rules for the example illustrated in Figure 1.2.

> *If current box contains a circle, change it into a diamond and move right.*
> *If current box contains an upward pointing triangle, change it into a circle and move right.*
> *If current box contains a diamond, change it into a circle and move left.*
> *If current box contains a downward pointing triangle, move right.*

The Turing machine illustrates several important concepts fundamental to the way we interact with modern computers. First, that programming of a computer involves reading and manipulating some kind of *data store* (the line of boxes containing symbols). We can get the computer to remember things for us which we can later recall when necessary. Second, to use a data store effectively, we must provide a series of *instructions* that tell the computer how to interact with the stored information. Those instructions must be unambiguously expressed and can both influence and be influenced by the contents of the data store.

Turing Machines and Abstraction

Alan Turing's conception of a machine that computes was made without any reference to how such machines would be built. It is a powerful example of *abstraction*, central to good object-oriented design, whereby we can remove the clutter associated with the mechanics of implementation in order to concentrate on *what* is done.

The fact that Turing's contribution to computer science was so profound and influential at a time when the available technology was capable of building only the most simple of machines is testament to the power of abstraction.

The nature of the instructions given to the computer varies widely in modern programming languages, but we can identify three distinct approaches. The first is known as *procedural* programming where the programmer supplies a sequence of instructions in some specified order. For example,

```
Step 1: Store the number 12 and refer to it as p.
Step 2: Store the number 25 and refer to it as q.
Step 3: Multiply p and q together and store the result as n.
Step 4: Display n on screen.
```

In procedural programming, the order of instructions is important (for example, if Steps 3 and 4 were swapped, the result of the process would be different). Languages like Fortran, C and Assembly Language are procedural and are often at the 'low-level' end of the programming continuum.

An alternative approach, known as *declarative* programming involves posing a problem rather than identifying the procedure for a solution. It involves two parts, namely (1) specifying some rules; and (2) posing a question. For example,

```
Rule 1: Green Park is connected to Oxford Circus via the Victoria Line
Rule 2: Oxford Circus is connected to Bond Street via the Central Line
Rule 3: Bond Street is connected to Marble Arch via the Central Line
Rule 4: Regent's Park is connected to Oxford Circus via the Bakerloo
        line.
Question: How many line changes are required to get from Marble Arch
to Regent's Park.
```

Introduction 7

In this case, the order in which we specify the rules does not affect the outcome of the process. We might also regard declarative programming as often a higher-level form of programming than the sequenced procedural programming. Examples of declarative languages include SQL (a database query language) and Prolog.

A third approach to programming involves elements of both procedural and declarative approaches and is known as object-oriented programming. The Java language is based on this form of programming. In order to appreciate the object-oriented aspects of Java programming, we should first consider the more general approach of object-oriented modelling.

1.4 OBJECT-ORIENTED MODELLING

Java is an object-oriented programming language. The meaning of this statement will probably only become clear as you progress through this book and you develop your own programming 'intuition'. Rather than giving a technical definition of this (rather overused) term let us make a few observations about object orientation in programming languages.

- An object-oriented programming language allows object-oriented models of the world to be more easily translated into computer instructions.
- Other examples of object-oriented languages include C++, Delphi, Smalltalk, Objective-C, Beta, Simula, CLOS and numerous others.
- It is possible to incorporate object-oriented ideas into virtually *any* programming language.

What these observations suggest is that the importance of object-oriented programming is not determined by the language, but by the underlying modelling of the world. A language like Java simply makes the process of implementing an object-oriented model a little less awkward than with some other languages.

So, what does object-oriented modelling involve? It should be regarded as a way of organising your thoughts when solving a problem. In particular, it involves breaking down a problem into smaller components, each of which have certain predictable behaviours and are able to share information with each other. This way of thinking may or may not come naturally to you, but hopefully by the time you get to the end of this book, you will be able to think in object-oriented terms without too much effort. As will be the case throughout this book, it is probably best illustrated with an example.

Suppose, that while sunbathing in your garden, you notice an entrance to an ants' nest. You observe that the ants that leave this nest appear to be exhibiting rather complex behaviour, spending some of their time exploring, taking food back to the nest along linear 'highways', interacting with each other and other insects in the garden. Being of an inquisitive mind you decide to model their behaviour in the hope of predicting how they might spread across your garden.

The object-oriented approach to this modelling problem suggests that you need to breakdown the problem into smaller components or *objects* that might be easier to model. The obvious unit to start with is the individual ant, and we might model its behaviour as follows:

Ant

Behaviour:

Choose a random direction and walk for 100 paces in that direction until either:

(i) food is found – in which case take food back to nest leaving a chemical trail

or

(ii) a chemical trail is found – in which case change direction and walk away from nest along trail.

Continue this behaviour until back in nest or ant has walked for 100000 paces without food.

This simple model is a rather poor predictor of the behaviour seen in the garden to say the least (Figure 1.3). However, we can make a more sophisticated model of the nest by simply adding lots of ants to it (Figure 1.4).

AntNest

Behaviour:

Populate a central spot with 1000 Ants over a 1 hour period.

There are two sets of advantages to this type of modelling approach.

1. *Simplicity*: The complex behaviour of a colony of ants has been made more manageable by breaking it down into more simple objects. The objects themselves in this example (ants and a nest) are relatively intuitive in that they correspond to real entities. If we (quite rightly) felt that a real ant is more complex than the model we have used, we might break it down further into 'smaller' objects such as an eating component, a walking component and a chemical communication component. One of the claimed advantages of object-oriented modelling is that it better represents the way in which we organise our own way of thinking.

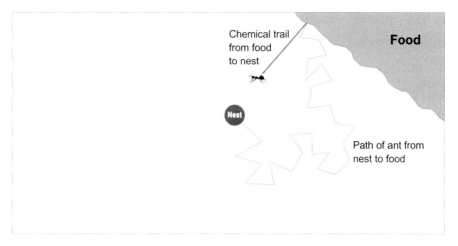

Figure 1.3 A simple ants' nest model containing only one ant.

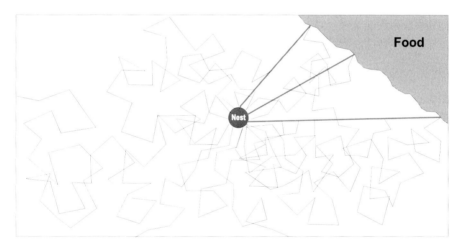

Figure 1.4 A More complex model with many ants.

2. *Extendibility*: Note how easy it was to go from our Ant model to the AntNest model, simply by adding the previously defined object. Suppose an entomologist friend of yours had spent the last 10 years developing a more sophisticated model of a single ant's behaviour. It will be easy to improve your nest model simply by replacing the Ant object with the new object. In doing so, we may not worry about the complex details of the model as long as we can observe its eating, walking and chemical trail laying behaviour. As we shall see in Chapter 3, this idea of hiding the details of objects (known as *encapsulation*) is an important and frequently used one. We may also extend our nest model by identifying different types of ant (e.g.

soldiers, workers and queen). We can use our existing model of the ant as the basis for these new classes, but then adapt it for each of the specialised types. This process (known as *inheritance*) is again widely used in object-oriented modelling.

We will revisit these ideas in Chapters 2 and 3, but for now you should begin to think about how any given problem can be broken down into self-contained 'objects', each of which address a simpler aspect of the problem.

1.5 WHY PROGRAM WITH JAVA?

By choosing to read this book, you have probably already made the decision that you wish to learn how to program with Java. However, it can be useful to consider some of the reasons why programming in general and Java in particular might be an appropriate route to take.

The discipline of learning any programming language tends to help in the understanding of computer-based environments generally, and GIS specifically. In learning a language, you will hopefully develop a 'way of thinking' that should help you to get computers to perform tasks, whether or not you actually end up programming in Java. This way of thinking involves taking often ambiguously defined problems and breaking them down into computable steps. Object-oriented modelling is one way of doing this, and is particularly suited to simplifying more complex tasks.

There are many object-oriented languages available, but a trend in the last decade or so has been for these languages to become increasingly complicated. They are often rather intimidating for the novice programmer. One of the design principles behind the development of Java was to try and simplify object-oriented modelling as much as possible. In comparison with C++, a similar object-oriented language, Java tends to be easier to use, especially for the beginning programmer.

Being a recently developed language, Java incorporates some of the more modern and useful developments in programming technologies. For example, Java is one of the few languages that has been designed at the outset to incorporate world-wide web technologies. Many of the early applications of Java involved creating *applets* – Java programs that could be placed on web pages and used by others over the internet. This association with the web has meant that recent web-friendly technologies such as the Extensible Markup Language (XML) and XML stylesheet language transformations (XSLT) have been rapidly integrated with Java.

Java is unusual as a compiled programming language in that it is *platform independent*. That is (in theory at least), you can write working programs in Java on a Windows computer, and someone else can modify and run the same program on a Unix or Apple Mac machine. The so-called '*write once, run anywhere*' principle has certainly proved attractive to application developers who wish to cut down on their development time on multiple platforms. This principle is particularly attractive to those producing internet-friendly programs, where it is not always possible to predict what computer a potential client might be using when running your code.

> **Java and JavaScript**
>
> There is sometimes confusion between Java and the completely separate language *JavaScript*. Java is a truly object-oriented language developed by Sun Microsystems since 1994. JavaScript is a dominantly procedural scripting language originally developed by Netscape to run inside web browsers. Some of the syntax of the two languages is similar, but the programming approaches that underlie each are very different. We will not consider JavaScript programming in this book.

1.6 A SHORT HISTORY OF JAVA AND THE INTERNET

Java is a modern language having been officially released to the world in 1995. This has the advantage of containing some of the most up-to-date ideas on programming (particularly communication between computers via the internet). It is probably fair to state that the core of the Java language is pretty stable now and has remained essentially unchanged for the last few years. However, as new technologies emerge, extensions and additional libraries of Java classes are continually added. This means that a typical Java installation is now much larger and feature-rich than it was two or three years ago.

To consider why Java has emerged today as a dominant development language, it is useful to consider briefly, the recent history of the programming languages that have influenced Java's development. In order to (over)simplify things, we might consider three distinct phases in the evolution of modern computer languages used for handling spatial information.

During the late 1970s and 1980s, the programming language C emerged as a standard for medium-level computer programming. During this period, many of those who used computers for handling spatial information were developing new data models and analytical processes. This type of development was illustrated by the first in the influential conference series *International Symposium on Spatial Data Handling* in Zurich, 1984. Much of the work reported there related to new structures and algorithms that improved the efficiency and effectiveness of the way computers were used to process spatial data. Most of it can also be regarded as *procedural* in nature, and languages such as C can be used to implement many of these ideas.

As hardware technology improved during the late 1980s and early 1990s, the use of computers tended to be much more graphically orientated. In particular, the metaphor of the windows-based 'desktop' emerged as the dominant working paradigm. The significance of this for programmers was that a linear procedural approach was poorly suited to a windowing environment where many processes would be occurring simultaneously, perhaps with interactions between processes in different windows. Programs in languages such as C became increasingly stretched and complicated when trying to deal with multiprocess event driven environment of the windowed desktop. Partly for this reason, object-oriented languages became more widely used. In particular, C++ emerged as an object-oriented replacement for C.

Since the early-1990s communication between computers via the internet has revolutionised their use. While languages like C and C++ can certainly be used to forge internet communication, programmers had to be exposed to some of the low-level complexities of networking. They were also confronted with the problem of programming for multiple platforms communicating over the same network. Partly for these reasons, Java was developed as an internet-friendly, platform independent programming language.

A History of Java Releases

The developers and owners of the official Java language are Sun Microsystems. They provided the first public '1.0a' release of the language in May 1995 along with an agreement with Netscape, that all new Netscape browsers would be able to interpret Java instructions.

For the next 18 months, 'Java 1.0' represented the standard form of the language upon which increasingly large number of web-based applets and applications were built.

In December, 1996, 'Java 1.1' was released with significant design enhancements and extra functionality over the 1.0 release. At that time, most browsers were not able to support the extra 1.1 functionality. For two years, further classes and enhancements were added to the 1.1 release. Web browsers begin to incorporate many of the Java 1.1 enhancements. The last release in this series was Java 1.1.8.

In December 1998, Java 2 was released. Somewhat confusingly, it was given the release number 1.2. Further enhancements continue to this day to be added to Java 2, with the current release at the time this book went to press being 1.4. In order to avoid the problems of continued changes to the language, Java was separated into (relatively stable) core functionality and (evolving) extension packages.

For more detail on the history of Java releases see also,
java.sun.com/features/1998/05/birthday.html for Sun's account of the early years of Java development
java.sun.com/features/2000/06/time-line.html for a graphical summary of the official Java development history.

Introduction 13

1.7 CREATING A WORKING PROGRAM

To develop your own Java programs, you need three pieces of software installed on your computer. You need some kind of *editor* that allows you to type in the text of a Java program. This might be a standard text editor that already exists on your computer such as Notepad on Windows or Pico on Unix. Alternatively, you might use an editor specifically designed for creating Java programs. Such editors tend to make the process of writing Java programs easier by using *syntax highlighting*. This involves automatically highlighting different parts of Java code in different colours.

Second, you will need to install a Java *Software Development Kit* (SDK) that contains the software that will translate the Java code you write into a form understandable by the computer. Sun and other vendors provide these for most platforms. Development kits are sometimes linked with editors and program management software in a single package called an *Integrated Development Environment* (IDE). Using an IDE can significantly speed up the process of designing, writing and testing your Java programs (see *Installing Java on Your Computer*).

Finally, you will also need a *Java Virtual Machine* in order to run your completed programs. A Virtual Machine (or VM) is included in the Java SDK from Sun as well as in most web browsers. Note that if you only wish to run Java programs written by someone else, this is the only one of the three components that you need installed on your computer. Since most computers have some web browser installed, this means that Java programs placed in web pages (known as *applets*), can be run by millions of users world-wide on a large range of computers.

Installing Java on Your Computer

If not already done so by someone else, to create working Java programs on your computer you will need to install some software first.

The Software Development Kit

The official releases of Java from Sun Microsystems come in two flavours – the Java Software Development Kit (SDK) and the Java Runtime Environment (JRE). In order to write your own programs, you will need a version of the SDK.

The SDK can be downloaded for free from Sun's Java website at www.javasoft.com You should follow links to 'Products and API's' and then the Java 2 Platform. This itself comes in two varieties – the standard edition and the enterprise edition. The standard edition should be sufficient for all the development and examples covered in this book.

> **Integrated Development Environments**
>
> To make the process of writing, compiling and testing your Java programs easier, you may also wish to consider installing an Integrated Development Environment (IDE). This piece of software groups together the editor required to enter Java code along with the compiler required to convert it into machine-readable instructions. Many IDEs also provide extra assistance such as project management, graphical user interface building and class design visualisation.
>
> There are many IDEs available, some of which are provided by commercial vendors, others of which can be downloaded from the web for free. Of particular use for the beginner is the IDE BlueJ, developed at Monash University, Australia. This freely downloadable package has been designed for teaching object-oriented concepts and is particularly helpful in its representation of Java objects and classes. BlueJ can be downloaded from www.bluej.org.
>
> Sun provide their own IDE – *Forte for Java*, a version of which can be downloaded from their site at www.sun.com/forte/ffj. This is a more complex IDE better suited to larger scale programming projects and can be a little confusing for the beginning programmer. For other commercial IDEs, the website *JavaWorld* provides a comprehensive listing and review:
> www.javaworld.com/javaworld/tools/jw-tools-ide.html

1.7.1 Writing, Compiling and Running

We will consider the process of creating a working Java program initially without reference to an IDE. Even if you intend to use an IDE in the future, it is useful, at least once, to be exposed to the three separate processes of editing, compiling and running a Java program from the 'command line'.

To create a set of instructions that the computer will understand, you need to do three things. The first stage is to type in the text of the Java program using an editor and store it somewhere on the computer. All Java programs you create in this way will be called `something.java` where `something` is the name you choose to give your Java program.

The second stage is to get the computer to translate (compile) the Java program you have typed into *bytecode*. Bytecode is a platform-independent set of instructions that is much closer to the machine code understood by computers.

To compile a Java program in this way you first need to open a command line interface to your computer. On windows this will be a 'DOS box' either accessible as a 'Start menu' option, or by selecting `Run...` from the Start Menu and typing the word `command`. In Unix, you need to start a shell of some kind. Once you have opened the relevant window (see Figure 1.5), you need to type the following to convert your Java code into bytecode.

Introduction 15

```
javac something.java
```

This will produce a new file in the same directory called `something.class` (known as a class file).

To get the program you have just created to work, you will then type:

```
java something
```

In each case, substituting the word *something* with the name you have given your Java program.

This runs your bytecode through the *Java Virtual Machine*, which does whatever you told it to do in your Java code. The next section should clarify this process by examining a real Java program.

Figure 1.5 Compiling and running a Java program on Windows and Unix platforms.

1.8 EXAMPLES

The code below shows a very simple Java program, called `Hello1.java`.

```
// *****************************************************************
/** Program to display a simple text message on the screen.
  * @author    Jo Wood.
  * @version   1.0, 12th September, 1997
  */
// *****************************************************************
public class Hello1
{
    // All java applications have a main() method.
    // This one displays a simple message.

    public static void main(String args[])
    {
        System.out.println("Java Programming for Spatial Sciences");
        System.out.println("==== =========== === ======= ========");
```

```
            System.out.println("");
            System.out.println("Program One: Hello World.");
    }
}
```

At this stage, you should note the following from this code:

- **Comments** Any code that is enclosed by the /* and */, or any code on a line after a // is ignored by the computer when translating into bytecode. These are shown in italics above.

    Despite being ignored by the computer comments are *very* important to us as they help to explain how the program works. All programs should start with a commented *header* explaining what the program does, who wrote it and when.

- **Classes** are the main building blocks in the Java language. This one is called Hello1 and its description is enclosed between { and } (known as *braces*). Note that all lines inside this class are indented by a few spaces.

- **main()** is the starting point for Java programs known as *applications*. As with the class definition, the description of main() is enclosed between braces and indented by a further few spaces.

- **System.out.println()** is the code that actually does something. It prints the message enclosed between quotes on the screen. Note that each of these lines ends in a semicolon.

We shall ignore the remaining parts of the program for the moment. The important point is that you begin to get a feel for what a Java program looks like.

To run this program, we will type the following at the command line:

```
javac Hello1.java
java Hello1
```

This will print the following on the screen:

```
Java Programming for Spatial Sciences
==== =========== === ======= ========

Program One: Hello World.
```

Finally, have a look at a more complex example that creates a graphics window and displays a message inside it.

```java
import java.awt.*;              // Needed to use graphics classes.
import javax.swing.*;

// ****************************************************************
/**
  * Hello2 - Program to create a graphics window and display a
  *          simple text message on the screen.
  * @author   Jo Wood.
  * @version  1.2, 25th August, 2001
  */
// ****************************************************************

public class Hello2 extends JFrame
{

    // All java applications have a main() method.
    // This one creates a graphics window.

    public static void main(String args[])
    {
        new Hello2("Java for Spatial Sciences");
    }


    // This method has the same name as the class (Hello2) and is
    // called a constructor. In this case it sets the window title.

    public Hello2(String windowTitle)
    {
        // Initialise window.
        super(windowTitle);
        setDefaultCloseOperation(WindowConstants.DISPOSE_ON_CLOSE);
        Container window = getContentPane();

        // Add two lines to the window.
        window.add(new JLabel("Program two:"),BorderLayout.NORTH);
        window.add(new JLabel("Hello again."),BorderLayout.CENTER);

        // Size window and make it visible.
        pack();
        setVisible(true);
    }

}
```

Much of this program, you will not understand the details of, but the important points to note are:

- Comments are used again to explain what the program does.

- Each time a *class* or a *method* is defined, its description is enclosed by braces and indented by a few spaces.

- As before there is a `main()` method, only this time we have defined a new one as well (`Hello2()`).

When this program is run (by typing `java Hello2`) a window similar to that shown in Figure 1.6 should appear.

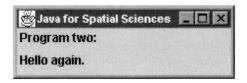

Figure 1.6 Output from `Hello2.java`

1.9 SUMMARY

In this introductory chapter, we have placed the Java programming language in the context of other approaches to computer programming. The idea of object-oriented modelling has been introduced emphasising that it allows the world to be modelled as a series of entities with their own state and behaviour. The concepts of *inheritance* and *encapsulation* have been introduced.

We have seen how Java is a modern and useful object-oriented programming language that allows the same program to be run on a variety of machines. In particular, a Java application consists of *comments* to make the program clearer, *classes* that form the building blocks of the language and *method*s that perform tasks. To run a program written in Java it first needs to be *compiled* into *bytecode* using the `javac` compiler, then run using the `java` interpreter.

By the end of this chapter, you should be able to

- give reasons why programming in Java is a useful process;
- identify advantages to the object-oriented approach to modelling;
- recognise a Java program;
- create, compile and run a simple Java program; and
- install the Java SDK and an IDE on you own computer.

You may wish to visit the companion website `www.soi.city.ac.uk/jpss` in order to test these learning outcomes.

CHAPTER TWO

Introducing Classes and Objects

In the last chapter, you were introduced to the process of *object-oriented modelling* which allows us to translate our models of the world into a form suitable for computer coding. We also saw how we might begin to go about writing simple Java programs using an editor, a compiler and virtual machine.

In this chapter, we shall look at the structure of a typical object-oriented Java program. In doing so, we shall develop the idea of object-oriented modelling and begin to relate it to the process of creating useful Java code.

2.1 CLASS DESIGN

We have already seen that it is possible to model an aspect of the world by breaking it down into a series of classes, each of which can have a certain behaviour. So, how do we apply that process to writing Java programs?

All working Java programs consist of one or more *class definitions*, in which we create, name and describe each class in terms of two fundamental characteristics – a class's *state* and *behaviour*. The state of a class is a list of properties it *has*, while a class's behaviour is a list of things it *does* (we will explore these two aspects in more detail below).

So, let us consider how we might go about creating some Java classes to model a particular problem. Suppose, we would like to write a Java program that models the use and maintenance of a local park. Our first task is to consider the entities that might make up our park model. Just as we might when designing a database schema, we should think of an entity as corresponding to a useful real-world object that has identifiable characteristics.

> *Before reading on, you might like to have a go at doing this yourself. Try jotting down on paper 4 or 5 classes, each of which represents a different aspect of the park model. Under each class, try to identify some of its properties in terms of things it has and things it does.*
>
> *Do not worry if this all feels a little abstract at this stage – this is your first attempt at designing an object-oriented model. You will certainly get better at it as we explore these ideas in this book.*

Figures 2.1 and 2.2 show a simple representation of a park both as a conventional map and as a *class diagram.*. The diagram shows that a Park class has been created that contains a GrassArea, a PlayArea, a FlowerArea, and a Path. In turn, the GrassArea also contains a Tree and a Path and the FlowerArea a Path.

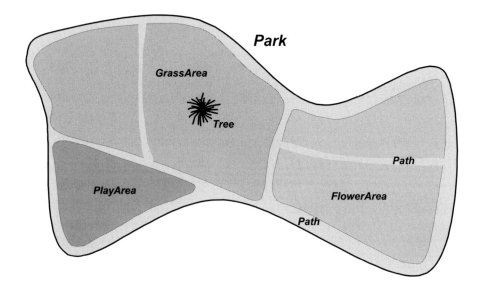

Figure 2.1 A map of a simple park.

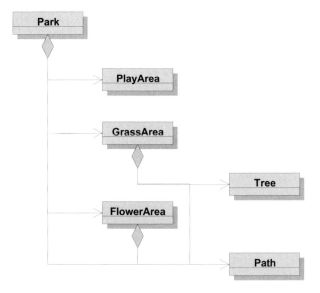

Figure 2.2 Class diagram representing the park.

Introducing Classes and Objects 21

The lines in the diagram indicate that one class 'contains' another; the rectangle at the diamond end of each arrow is the container class while the rectangle at the arrow end is the contained class. This form of relationship between classes is sometimes known as *composition* or a '*has a*' relationship (e.g. 'FlowerArea has a Path'). It allows us to build up more complex classes based on groups of simpler classes.

It is good practice when creating a new program to think carefully about the class design before you start coding. Sketching out a class diagram provides an easy way to see the 'big picture' and experiment with different forms of object-oriented model. Time spent thinking about sensible class design will almost certainly save time and effort when you get to coding and debugging your Java programs.

UML Diagrams

The class diagrams used in this book are simplified versions of what are known as *UML Diagrams*. The Universal Modelling Language (UML) is powerful language-independent set of conventions for describing systems and processes. It allows designers to specify the details of an object-oriented model, the state and behaviour of classes and the way in which they interact with one another. A well-specified UML description can be translated with relative ease into any object-oriented language such as Java or C++.

UML diagrams use a set of graphical conventions to indicate the relationship between classes in an object-oriented model. Typical models can appear somewhat complicated when fully specified as a UML diagram and can be rather intimidating for the novice programmer.

For example, UML makes the distinction between two types of 'has a' relationship, namely *composition* and *aggregation* each with their own symbol. For this reason, we have adopted a simplified set of conventions in this book.

For details of the full UML specification, see www.uml.org.

Note that in this example, when one class 'contains' another, it happens to do so in a spatial sense. The Tree really is contained within the GrassArea for example. In many other cases in Java programming, that relationship will not necessarily be so spatial. For example, we might add a new class to our Park model called Management that might contain a further class called ParkKeeper. This form of composition is more conceptual than spatial, but equally valid.

Our Java program that modelled parks would contain six classes, one for each of the orange rectangles in the class diagram in Figure 2.2. Each class would contain information about its state and behaviour. So, for example, the Tree class might be defined as follows:

```
Tree
─────────────────────────────
State:
  Tree type
  Tree height
  Age
  Location
─────────────────────────────
Behaviour:
  Grows by 5% per year.
  Sheds leaves every Autumn.
```

2.2 CLASSES AND OBJECTS

We shall look at how we might code a class's state and behaviour in the next section, but first we must make an important distinction between the idea of the *class* and the *object*.

You may have noticed that despite Java being an 'object-oriented language' we have been modelling entities as classes not objects. So, what is the difference between the two, and when do we use objects?

When we design an object-oriented model, what we actually do initially is try to create generic classes of things rather than specific instances of things. So, for example, in the example above we created a `Tree` class that can be used for representing *any* tree. We can distinguish that from an object that might represent a specific *instance* of a tree, such as the one below.

```
oakTree1 is an instance of Tree
─────────────────────────────
State:
  Type: Oak tree
  Height: 17.5m
  Age: 55 years
  Location: (44500,478240)
─────────────────────────────
Behaviour:
  Will grow 87cm this year.
  Will shed its leaves in 5 months time.
```

Separating the generic (class) from the specific (object) is one of several characteristics of object-oriented modelling that makes it such a useful and powerful process. Once we have created generic classes of things, we can reuse them in a variety of ways. For example, although our particular park contains three distinct sets of paths (one separating the areas of the park, one in the flower area

Introducing Classes and Objects 23

and one in the grass area), which would each be modelled as an object, they are all based on the same shared `Path` class.

> **Objects, Classes and Relational Databases**
>
> For those who are familiar with relational database design, there is an analogy here with the creation of the tables used to represent information. Creating tables and identifying the fields (columns) within them is equivalent to creating a class in object-oriented modelling. Populating the records (rows) within a table is equivalent to creating objects out of the class, where one row is analogous to one object. You might also notice a similarity between our class diagrams and *entity-relationship diagrams* that allow us to visualise the connections between relational tables.
>
> Unlike relational tables, objects are not only capable of storing data, but are also able to enact certain *behaviours*, in other words, process and manipulate those data too.

So, let us turn now from the theory of creating classes and objects to the practice of doing so in Java.

2.3 THE STATE OF A CLASS

Remember the state of a class is a description of the information it could contain (tree height, age, location, etc. in the example above). In this section, we will consider how we get Java to store such information using the concept of the *variable*.

For those who have ever programmed in other languages, the concept of the variable is likely to be a familiar one. But as we shall see, in Java we can store a range of types of variables from simple numbers to complex objects.

2.3.1 Simple Variables

A variable is just a portion of the computer's memory set aside to store a number or series of numbers. In Java, variables are sometimes referred to as *fields*. It is analogous to a pocket calculator's M (memory) button, which allows a number to be stored somewhere in its memory.

As will become clearer later on, it is useful to think of the computer's memory as one long conveyor-belt upon which sit boxes with names containing numbers. At any time, we can open the box and examine or change its contents. All variables have names to identify them, which can be anything we want within certain limits.

Have a look at the following program as an example of how variables can be used in a Java program.

```java
// ****************************************************************
/**
 * Pythag - Program to demonstrate Pythogoras' theorem for right
 *          angled triangles. It demonstrates the use of simple
 *          variables (fields).
 * @author   Jo Wood.
 * @version  1.1, 14th May, 2001
 */
// ****************************************************************

public class Pythag
{
    // All Java applications have a main() method.
    // This one does a simple Pythagorean calculation.

    public static void main(String args[])
    {
        // Declare fields and initialise values (input).
        int firstNum = 17;
        int secondNum = 6;
        int answer;

        // Do the calculation (process).
        answer = (firstNum*firstNum) + (secondNum*secondNum);

        // Display the results (output).
        System.out.println("The square of the hypotenuse is "+
                           "equal to the sum of the squares "+
                           "of the other two sides.");
        System.out.println("For example...");
        System.out.println(answer+" = "+firstNumber+
                           " squared + "+secondNumber+" squared.");
    }
}
```

Compiling and running this program will produce the following output.

```
The square of the hypotenuse is equal to the sum of the
squares of the other two sides.
For example...
325 = 17 squared + 6 squared.
```

As with all Java programs, we have started by defining a class, which in this case is called `Pythag`. Everything inside this class definition is indented by a few spaces. Again, as with all other Java applications we have seen, there is a `main()` method. This is the start point for our program.

Comments are used to make the program clearer to read. The header at the top explains what the program as a whole does, the other comments explain each element. The input, process and output elements are emphasised in this case.

Introducing Classes and Objects 25

The input element involves declaring and initialising three variables (see below). Each of these variables will store a single whole number (integer). The number 17 is placed in the variable called firstNumber, 6 is placed in the variable called secondNumber, while the variable called answer is left empty.

The process element of the program involves a simple calculation. The number stored in the variable firstNumber is multiplied by itself (the * means multiply) and then added to the contents of secondNumber multiplied by itself. The result of this calculation is placed in the variable answer.

The output element of the class displays the results of the calculation on the screen. It uses the line System.out.println() which we have seen before, but this time in a more sophisticated way. If we wish to print the contents of the variables we have defined, they must be included in the println() call outside of the quotation marks. If they are to be combined with other text, they are separated by the + symbol.

Note that what we have described inside the main() method is essentially *procedural*. That is, we have issued a series of instructions, which are executed by the computer line by line moving from top to bottom through the code. As we shall see when we develop some more object-oriented ideas, program flow is not always this linear (or straightforward).

2.3.2 Naming Variables

Unlike a calculator, we can give variables more explanatory names than M. When naming a variable you should use a name that explains what it will hold. Do not make variable names too long or too short.

The following are all legal and sensible names for variables:

```
answer1
pi
roadWidth
```

The following are all illegal variable names:

```
1answer           (must not start with a number)
twenty-two/seven  (must not contain mathematical operators)
road width        (must not contain spaces)
```

As a general rule, the following are legal but poor names for variables:

```
b                 (what does this represent?)
variable          (again, what does this represent)
the_number_of_people_included_in_survey  (a little cumbersome!)
SUMMARY           (by convention most variables are named in lower case)
```

As you may have already found out, all Java programs (and program names) are case sensitive. That is, Java would treat `answer` and `Answer` as completely different names. Although this might seem unnecessarily pedantic, it is useful for distinguishing certain types of variables and classes in Java. Here are a few of the naming conventions (more will follow later on).

- All variables and methods should be in lower-case. Where a name consists of more than one word, the first letter of each subsequent word should be capitalised (sometimes known as intercapping). So, for example:

    ```
    roadWidth
    distanceToCentre
    setXYPosition
    ```

- All classes should be lower case with an initial capital letter (as `Pythag` above). Note also that the file in which the class is defined should also have the same name as the class (but with a `.java` at the end).

Abbreviated Variable Names

Over the years since Java has been widely used, there have developed a number of conventions for naming variables that appear to contradict the principle of explanatory naming. In particular, there are some processes that are sufficiently common that abbreviated names are often used. Table 2.1 shows some examples (do not worry if their context does not make sense to you yet – they will be covered later in this book).

Table 2.1 Examples of abbreviated variable names

Example	Explanation
`for (int i=0; i<100; i++)`	Simple counters in loops (see Section 4.4 below) often use the letter `i` when its only function is to count the progress through a loop.
`paintComponent(Graphics g)`	A graphics context object (see Section 4.7 below) is usually given the name `g`.
`catch (Exception e)`	Exceptions are often represented with an object named `e` or `ex`.
`Iterator i=collection.iterator();`	An iterator counts through items in a list of objects. Frequently, this is given the letter `i`, just as it is when used in a loop.

2.3.3 Types of Variables

Different types of number take up different amounts of computer memory. As you might expect, `age=5;` takes up less memory than `fgnbm=4.669372418117;` To be as efficient as possible, we must *declare* all variables in our programs, telling the computer what sort of numbers are likely to be stored within them. Variable declaration is usually carried out at the beginning of a method (as in `Pythag.java` above).

There are two basic types of number variables: (1) whole numbers; and (2) real numbers. Each of these has several types depending on how large (or small) the numbers likely to be stored in them (in addition, both logical and text characters can also be stored as primitive variables).

Table 2.2 Java primitives

Name	Bits	Minimum	Maximum	Type of number
byte	8	–128	127	whole number
short	16	–32768	32767	whole number
int	32	-2^{31}	$2^{31} - 1$	whole number
long	64	-2^{63}	$2^{63} - 1$	whole number
float	32	very small	very large	real to ~7 decimal places
double	64	even smaller	even larger	real to ~15 decimal places
boolean	1	(false) 0	(true) 1	Binary
char	2	(encoding dependent) 0	(encoding dependent) 65535	Unicode text character

The commonest type of whole number variable is the `int` (integer) type, which is suitable for storing typical whole numbers. If for some reason, you need a variable to hold a larger number than $2^{31} - 1$, then the `long` type can be used. If you have reason to save memory and know that only small whole numbers are to be stored, you can use the `short` or `byte` variable types. For variables that need only store a `true` (1) or `false` (0) value, the `boolean` type can be used.

The commonest type of real number variable (i.e. one that holds a number with a decimal point) is the `float` type. This can hold numbers with a precision of about seven decimal places. If you know you will require greater precision than that, you can declare `double` (precision) variables.

A variable is *declared* as

`variable_type variable_name;`

For example,

```
int score;
double average;
```

Additionally, variables should also be *initialised*. That is, they should have some initial value placed inside them. It is good programming practice to declare variables and where possible initialise them at the beginning of a method or class.

A variable is initialised by including the line in the form:
```
variable_name = value;
```

For example,

```
score=0;
average=50.0;
```

It is also possible to combine the process of initialising and declaring a variable in a single line.

```
variable_type variable_name = value;
```

For example,

```
int score=0;
double average=50.0;
```

It is important to be clear on the difference between declaring something where we simply state what type of information could be stored in a variable, and initialising something where we place some data inside a variable. This distinction will become more important when we start manipulating classes and objects.

Primitives and Classes

The number variables we have described are known as *Java primitives* since they store the building blocks of any computer program – numbers of various sizes. In addition to each of these primitives, Java can also model numbers as classes. There exists a class for each primitive and is given a similar name, but with an upper-case initial letter (`Boolean`, `Character`, `Byte`, `Short`, `Integer`, `Long`, `Float` and `Double`).

For basic numerical calculations, like the ones we have seen so far, it is more efficient to use primitives to store numbers. However, as we shall see, the class equivalents also have certain useful behaviours that allow us to do things like finding the largest and smallest storable numbers and converting numbers from one class to another.

2.3.4 Typecasting

Java is known as a *strictly typed language*. This means that it is reluctant to let you assign variables of one type to another. For example, continuing the example above, if you attempt to compile the line

```
score = average;
```

Java will try to place the number stored in the variable `average` into the variable `score`. In doing so, the Java compiler (`javac`) will report the error

```
Incompatible type for =. Explicit cast needed to
convert double to int.
```

Although this may seem inconvenient, it will generally be bad practice to try to assign a double precision number to an integer, as all the digits after the decimal point will be lost in the process. However, the line

```
average = score;
```

will not generate a compiler error. There should be no loss of precision in converting an integer to a double precision number, and the new value for `average` will now be `0.000000000000000`.

A slightly more subtle version of this problem can arise when we assign numbers to variables. This is because whenever we use a number in a program, the Java compiler implicitly assumes it is an integer if it contains no decimal point, or a double precision number if it has one. Therefore, the following line will produce a compiler error:

```
float temperature = 25.3;
```

The error is produced because Java tries to assign `25.3`, which it assumes to be a double precision number, to a floating point variable. The way round this is to use what is called *typecasting*. This involves explicitly changing one type of number or variable into another. It should always be done with caution since it can involve a loss of precision.

To typecast a variable, simply place the new type of variable in brackets in front of the variable/number to be changed. For example,

```
score = (int) average;
float temperature = (float)25.3;
```

Because the typecasting of numbers is more common than variables, Java includes a shorthand notation for changing the way numbers are represented. To force a number to be represented as floating point, place an `f` immediately after it. To represent it as a double precision number, place a `d` after it. For example,

```
float temperature = 25.3f;
double average = 0d;
```

To consolidate, take a look at the following program:

```
// ******************************************************************
/** Program to illustrate the input, process and output elements
  * of a program. It also shows how variables of different types
  * may be declared, initialised and used.
  * @author Jo Wood.
  * @Version 1.1, 14th May, 2001
  **/
// ******************************************************************

public class Acidity
{
    // All java applications have a main() method.
    // This one does a simple mean and difference calculation.

    public static void main(String args[])
    {
        // Declare and initialise variables (input).
        int easting  = 329440;       // National Grid coordinates
        int northing = 441660;       // of sample point.
        int sampleNum = 12;          // Sample point number.

        double acidityNew = 6.53;    // Soil pH at sample.
        double acidityOld = 6.27;    // Previous pH reading.
        float change;                // Change in pH value.
        float average;               // Average pH value.

        // Do the calculation (process).
        change  = (float)(acidityNew - acidityOld);
        average = (float)((acidityNew + acidityOld)/2.0);

        // Display the results (output).
        System.out.println("\n\nSoil quality monitoring.");
        System.out.println("==== ======= ===========");
        System.out.print   ("\tFor sample " + sampleNum + " at (");
        System.out.println(easting + "," + northing + ")");
        System.out.println("\tthe average pH value is " + average);
        System.out.println("\tChange since the last sample is "+
                             change);
    }
}
```

As with all programs, we start off by explaining what the program does in a commented header. The variables representing the location, sample number and acidity readings are all declared and initialised. The two variables that will store the average and change in acidity are then declared. The calculation of change and average involve processing double precision variables but storing the result as floating point, so we are forced to typecast the result of the calculation in order to fit it inside the two floating point variables. You may also notice that the output section of the code involves more sophisticated formatting, which we will cover in more detail in the following section.

2.3.5 Displaying Variables on Screen

The program above produces output as follows:

```
Soil quality monitoring.
==== ======= ===========
    For sample 12 at  (329440,441660)
    the average pH value is 6.4
    Change since the last sample is 0.26
```

Text can be combined with the contents of variables simply by adding the variable to the list of things to be printed i.e. inside the brackets of `println()` but outside the quotes. If a `\n` is included within the quotes, a new line is thrown rather than the characters `\` and `n` being displayed. This represents one of several special *escape codes* that may be used to control output. The other one seen in this program is the `\t` code, which indents subsequent text by one tab stop (about 8 spaces).

We can control positioning of text using special characters that begin with the `\` symbol.

```
\n      new line
\t      tab stop
\b      backspace
```

Suppose we wish to print a `\` on the screen. How will the computer know that it is not part of a special positioning character? We simply place a `\` before the character we wish to print. So therefore,

```
\\      outputs a single \
\"      outputs a quotation mark.
```

2.4 THE BEHAVIOUR OF A CLASS

So far, we have concentrated on how we might get a Java program to store items of data by using variables that hold numerical information. Such variables help to define the *state* of an object. We have also seen that classes and objects do not just store information, but they can be given particular *behaviours*. In this section, we will look at how we can begin to program object behaviour using the concept of the *method*.

2.4.1 Methods

As we have already seen, a method in the Java language is a section of a class that does something. A method is identified with a unique name (just as variables are) and is followed by brackets `()`. The method itself is enclosed in braces `{}`. So far, all our programs have had at least one method called `main()` which is always the first method called by a Java application.

When designing a class, a method should correspond to an identifiable aspect of the class's behaviour. So, for example, if we consider our `Tree` class that we identified earlier on in this chapter, we might create a method to represent the growing behaviour of a tree. We can then code our `Tree` class as follows:

```
// ****************************************************************
/** Class for representing trees.
  * @author   Jo Wood
  * @version 1.0, 14th May, 2001
  */
// ****************************************************************

public class Tree
{
    // ------------------ Object variables ---------------------

    private float height;        // Stores height of the tree.

    // -------------------- Constructor ----------------------

    /** Creates a new tree and gives it an initial height.
      */
    public Tree()
    {
       height = 1;         // Set initial height at 1m.
    }

    // ---------------------- Methods -------------------------

    /** Allows the tree to grow by 5%.
      */
    public void grow()
    {
       height = height + (0.05f * height);
    }
}
```

As it stands, this program will not do much since we do not have a main() method, and the class needs to be linked with the other classes that make up a Park. However, it does illustrate two important ways in which we can use methods.

The class consists of two methods, each of which manipulates the variable height in a simple way. The first method, called Tree is a special type of method called a *constructor*. For the moment, we can regard a constructor as a method that shares the same name as the class itself (Tree) and is called automatically whenever an object is created from a Tree class. In this example, the constructor simply initialises the variable height with a value of 1m.

The second method, called grow() will not be called automatically, and as we shall see in the next chapter, must be invoked explicitly somewhere else in our Java program. If it were to be invoked however, it would find the current value of height and increase it by 5%.

Do not worry about the words `private`, `public` and `void`. These will be explained in the next chapter. For the moment, what you should understand is that we can add many methods to a class to get it to perform a behaviour of some kind. Each will have a broadly similar form to the ones shown above.

2.5 MAKING CODE CLEAR

You will recall from the first chapter that the objective of any programming language is to bridge the gap between our way of modelling the world and the computer's own binary representation. Consequently, a good program (in any language) should achieve two things. First it should provide the unambiguous instructions that the computer requires in order for it to run the program. Second, the program should communicate to *us* what it is trying to achieve. In order to fulfil the second of these objectives, all Java programs should be easy to read and well documented. From the outset of your programming in Java, you should get into the habit of clear layout and thorough documentation.

2.5.1 Use Sensible Names

All variables, classes and objects should be given sensible names that describe what each does. These names should be concise and unambiguous as possible. By convention, all variable, object and method names are given in lower-case letters with 'intercapping' where necessary. Class names are also intercapped, but should also start with an initial upper-case letter.

2.5.2 Indent Code

Java programs tend to be composed of blocks of code that make up classes, methods, and as we shall see in Chapters 4 and 5, loops and conditional statements. Each of these tends to be enclosed in *braces* { }. The braces that enclose a block of code should always be vertically aligned making it easier to identify visually, the start and end of a given block. All code within a set of braces should be indented by several spaces or a tab.

For example,

```
public class AnyClass
{
    // Code here is indented.

    public void anyMethod()
    {
        // Code here is indented further.
    }
}
```

> **Alternative Layout Conventions**
>
> There is an alternative convention of indenting and positioning of braces that you may well see in other people's code. This convention places the opening brace of a block of code at the end of the preceding line, so that the example above would look like this:
>
> ```
> public class AnyClass {
>     // Code here is indented.
>
>     public void anyMethod() {
>         // Code here is indented further.
>     }
> }
> ```
>
> This has the advantage of saving some space on the page or screen, but makes it harder to spot the structure of a program from its visual appearance.

2.5.3 Document Code

Comments are vital when producing good understandable program code. You should get into the habit of placing comments in your programs *as you write them*, not as an afterthought when the program has been written. By commenting as you program, you will find program development much easier and will be less likely to make mistakes.

Java uses three types of commenting.

```
// Single line comments
```

These are used for short comments that might explain a small part of a program, or perhaps explain what a variable will store.

```
/* Multi-line comments can be used when
   a slightly larger explanation is required.*/
```

This might include descriptions such as how to use the program, its copyright or development history.

```
/** Javadoc comments are special versions of
    the multi-line comment.*/
```

They are identified by having at least two ** at the opening of the comment. They should be used when creating a class header and explaining what a method does. We shall see in Chapter 6 why these are very important in the documentation process.

Introducing Classes and Objects 35

Commenting code should be seen as part of the programming process. Usually when writing a Java program, the first thing you will do is to design the class structure (as in our `Park` example above). Once a class has been created, its methods and fields are then likely to be identified. You can do this by creating a 'skeleton' class containing just the name of the class, any likely fields and methods. Each of these should be commented before you get down to the work of actually adding any further Java code. In this, way it becomes easier to maintain a link between the abstract class design and the practice of building up Java code.

2.6 DISTINGUISHING CLASSES FROM OBJECTS IN JAVA

Once we have created a general class in Java, it becomes very easy to create any number of objects out of that class. So, how do we create an object in the Java language?

To create an object from a class that has been defined somewhere in our Java program, we use the special keyword `new`.

The general form of object creation is as follows:

```
className objectName = new className();
```

For example,

```
Tree oakTree = new Tree();
Tree elmTree = new Tree();
```

This process is sometimes called *instantiating* an object from a class. In other words, the object is regarded as an *instance* of the class from which it was created.

Note that keeping to the convention of naming classes with an initial capital letter and objects with a lower-case letter (as we do for primitive variables) helps us to distinguish between general classes and specific objects.

Once we have created an object in our Java program, we can invoke any of the methods that may have been created when designing the class. To call a method that is part of an object we have created, we name the object and method separated by a . (dot).

The general form used to invoke an object's method is as follows:

```
objectName.methodName();
```

For example,

```
oakTree.grow();
```

This process is best illustrated with our Park example shown in Figure 2.3.

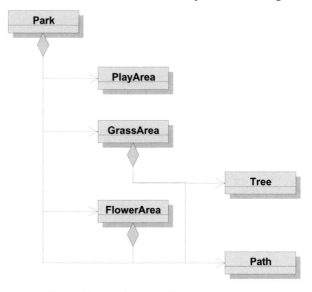

Figure 2.3 Class diagram of the park model.

Just to remind ourselves, the class diagram above represents our simple park model. The Park is *composed* of a PlayArea, GrassArea, FlowerArea and Path. The GrassArea itself contains both Tree and Path objects.

The code below shows how we describe the composition of the Park class in Java.

```
//    ****************************************************************
/** Simple model of a Park containing vegetation and recreational
  * facilities.
  * @author    Jo Wood.
  * @version   1.2, 27th August, 2001
  */
//    ****************************************************************
public class Park
{
    // ------------------- Object variables --------------------

    private GrassArea grass;          // Grass
    private PlayArea play;            // Swings and slides
    private FlowerArea flowers;       // Flower bed.
    private Path path;                // Path running through park.


    // --------------------- Constructor ----------------------

    /** Creates the park with its plants and entertainments.
      */
    public Park()
```

Introducing Classes and Objects 37

```
    {
        // Create new objects (composition).
        grass   = new GrassArea();
        play    = new PlayArea();
        flowers = new FlowerArea();
        path = new Path();

        grass.grow();
    }
}
```

The structure of this class is pretty similar to the ones we have seen already. After a commented header, we declare the name of the class and identify its state. In this case, the state of the park is represented not by simple number variables as we have seen previously, but four *objects* (`grass`, `play`, `flowers` and `path`).

The constructor, which you will recall, is simply a special method that shares the same name as the class in which it sits, initialises each of the four objects using the keyword `new`. Each of the classes that represents things within the park has a similar structure. For example, we might create the `GrassArea` class as follows:

```
// ****************************************************************
/** A simple model of a patch of grass.
  * @author    Jo Wood.
  * @version   1.2, 27th August, 2001
  */
// ****************************************************************
public class GrassArea
{
    // ------------------ Object variables --------------------

    private Tree oakTree;          // A lovely old oak tree.
    private Path grassyPath;       // Path running through park.

    // -------------------- Constructor -----------------------

    /** Creates a grassy area complete with path and tree.
      */
    public GrassArea()
    {
        oakTree = new Tree();
        grassyPath = new Path();
    }

    // ----------------------- Methods ------------------------

    /** Grows the grass and any other features of the grass area.
      */
    public void grow()
    {
        oakTree.grow();
        oakTree.displayDetails();
    }
}
```

The GrassArea class contains a method grow() in addition to its constructor, that in this case, forces the oak tree within it to grow. Note that to 'call' a method (i.e. force the behaviour represented by the method to be enacted) we use name the object and the method separated by a dot.

The two methods inside Tree are shown in the code for the Tree class below:

```java
// ****************************************************************
/** Represents a Tree with its own height, age and location.
  * @author    Jo Wood
  * @version   1.1, 23rd May, 2001.
  */
// ****************************************************************

public class Tree
{
    // ------------------- Object variables --------------------

    private float height;         // Stores height of tree.
    private int   age;            // Age of tree.
                                  // Coordinates of location of tree.
    private int   locationX,locationY;

    // -------------------- Constructor -----------------------

    /** Creates a new tree and initialises its state.
      */
    public Tree()
    {
        height = 1;
        age    = 2;
        locationX = 0;
        locationY = 0;

        displayDetails();
    }

    // ---------------------- Methods -------------------------

    /** Grows the tree by 5% and ages it by a year.
      */
    public void grow()
    {
        height = height + (0.05f * height);
        age = age + 1;
    }

    /** Displays details of the tree.
      */
    public void displayDetails()
    {
        System.out.println("Tree is " + age + " years old, ");
        System.out.print(height + "m tall and found at (");
        System.out.println(locationX + "," + locationY + ")");
    }
}
```

2.6.1 Constructors and Program Flow

So, what will happen if we try to 'run' the program above? The answer is that so far we will not be able to run the program in the conventional sense because we have no class with a `main()` method inside it.

We can create a new class with its own `main()` method that simply creates a new `Park()` object:

```
// ****************************************************************
/** Very simple class that starts the Park program by creating a
  * new Park object.
  * @author Jo Wood.
  * @version 1.1. 23rd May, 2001
  */
// ****************************************************************
public class RunPark
{
    /** Creates a new Park object.
      */
    public static void main(String args[])
    {
        Park myPark = new Park();
    }
}
```

The `main()` method creates a new object from the `Park` class just as we have seen in the other methods above.

By now, you will have noticed that the 'program flow', in other words, the order in which Java commands are invoked, is getting a little more complicated.

> *Before reading on, look at the four classes above and try to work out what will happen when the entire program is run. What do you think will appear on the screen once the `main()` method has been called?*

The order in which the elements of the code are invoked is as follows

1. The `main()` method inside `RunPark` is called and creates a new `Park` object.
2. The constructor of `Park` is automatically called which in turn creates a new object out of the `GrassArea` class.
3. The `grass` object then creates new objects out of `Tree` and `Path` (`oakTree` and `grassyPath` respectively).
4. The constructor of `Tree` calls its `displayDetails()` method which reports the tree's age, height and location.
5. The remaining lines of the `Park` constructor are then executed which create new objects out `PlayArea`, `FlowerArea` and `Path`.

6   Finally, the last line of the `Park` constructor invokes the `grow()` method of `GrassArea` which grows the oak tree and reports its new age, height and location.

Note that the order in which methods are defined inside a class is irrelevant to the order in which they are invoked. It is only when they are called from an object that the method is actually enacted.

Therefore, the output to the screen after calling `RunPark` will be:

```
Tree is 2 years old,
1.0m tall and found at (0,0)
Tree is 3 years old,
1.05m tall and found at (0,0)
```

You should be able to deduce from this that a constructor is a method that is automatically called whenever an object is created from a class. We do not have to call it explicitly like we do with other methods, and it provides us with an opportunity to initialise any variables in our object and generally perform any actions that we know in advance must be taken before other methods are called. We shall see how we can make use of constructors more fully in Chapter 3.

If you have been paying a great deal of attention, you may have noticed that our `RunPark` class does not seem to follow quite the same rules as our other classes and objects. Nowhere does there appear to be a line that creates an object out of `RunPark`. In fact, in all the classes that have contained a `main()` method, we appear to be missing an object instantiation.

The reason for this relates to the (as yet unknown) use of the word `static` in front of the `main()` method. We will explore the term `static` more fully in Chapter 6, but for the moment, you can regard it as a term that allows us to call a method or declare a variable directly from a *class* rather than an *object*. As a consequence, there is no need to create an object from `RunPark` as the `main()` method is called directly from the `RunPark` class.

2.7 SUMMARY

In this chapter, we have continued to develop the idea of object-oriented modelling by making a distinction between generic classes (for example, any tree) and specific instances of objects (for example, that oak tree over there). We have also seen how object-oriented models are likely to be made up of several classes, each of which are connected in some way.

We have begun to see how we can relate the abstract process of object-oriented modelling to the specific process of creating Java programs. We have seen how each class in a design can be created in Java, with its own state represented by variables and behaviour represented as methods.

We have seen how we can begin to manipulate the contents of variables in Java and display the contents of those variables using simple screen output. We have also considered how methods defining the behavioural aspects of a class can be invoked, either automatically as a constructor or explicitly using an `object.method()` call.

By the end of this chapter and associated exercises on the web, you should be able to

- design a simple object-oriented model made up of several linked classes;
- understand what composition is and why it is useful in designing classes;
- understand the difference between a class and an object;
- create a simple class in Java containing variables and simple methods;
- apply simple arithmetical manipulation to variables and display the results on screen; and
- lay out Java code to be clear and readable.

CHAPTER THREE

Developing Classes and Objects

Building object-oriented models of the real world not only involves identifying the classes that make up the model and their associated state and behaviour. We must also provide connections between these classes to form a linked model. In this chapter, we consider some of the mechanisms we can use to link classes and objects building on the ideas covered in the previous chapter. In doing so, we will be able to increase greatly, the power and flexibility of the object-oriented modelling process. These ideas are illustrated using Java's own graphics classes and in the initial design of an 'ants in the garden' model.

3.1 INHERITANCE

From the outset of this book, it has been suggested that one of the goals of object-oriented modelling is to create independent classes of things, each of which have their own autonomous states and behaviours. But, it should also have become apparent that classes that are entirely independent of others in the same model are pretty much useless. For example, in our 'Ants in the Garden' model, there would not have been much to gain if we had added a `GardenTable` class if it had no effect on the ants and was not itself affected by ant behaviour.

In other words, our models tend to consist of classes and objects that in some way can be related to one another. There are many ways in which this can be done, and we have already seen one of those – namely *composition* where one class or object contains one or more other objects (for example, the `Park` class developed in the previous chapter was composed of objects made from `GrassArea`, `Tree` etc.). This was characterised as a '*has a*' relationship.

The second important way in which we can relate classes together is via *inheritance*. This process allows us to create a new class, not from scratch, but by using an existing class as a 'template'. The new class inherits all the methods and object variables of the 'parent' class, but can have its own additional state and behaviour. This allows us to group similar classes together in a *hierarchy* with more specialised classes inheriting the characteristics of more general ones. For example, we might first create a class called `Ant` and then create three new classes `WorkerAnt`, `SoldierAnt` and `Queen`, all of which inherit `Ant`'s characteristics but with further specialisms. As a parallel to the '*has a*' relationship of composition, inheritance is sometimes characterised as an '*is a*' relationship.

Inheritance is useful in object-oriented modelling and Java programming because many features in the real world can show similar hierarchical organisation. In practice, we tend to use inheritance in one of two ways, known as *extension* and *generalisation*. Extension is the process of identifying an existing class that contains much of the functionality we might wish to use, creating a 'subclass' from it and adding extra functionality as needed.

Generalisation involves approaching the hierarchy from the opposite direction whereby we identify the functionality that two or more classes have in common, remove that functionality and place it in its own class and allow our original classes to inherit it. This process allows us to take an existing object-oriented model and improve its efficiency by removing duplication in class design. When you build your own models, you should think carefully about the generalisation process as it can often yield significant efficiency and reusability benefits.

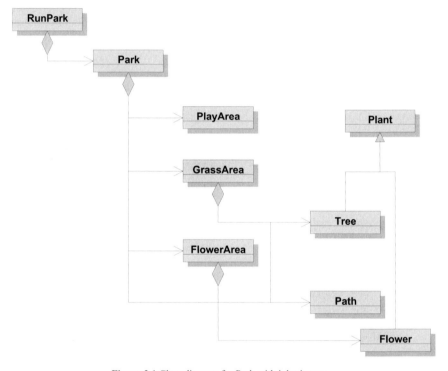

Figure 3.1 Class diagram for Park with inheritance.

Let us consider how we might use inheritance in our Park model developed in the previous chapter. One obvious candidate for inheritance generalisation is our model of vegetation in the park. It contains trees, flowers and grass, all of which have certain characteristics in common. It makes sense to identify this common element and place it in its own Plant class. Both Trees and Flowers are

Developing Classes and Objects

examples of `Plants`, but may have further tree-specific or flower-specific behaviour.

Figure 3.1 shows a possible structure for our park model using inheritance. Note that the convention in our class diagram for representing inheritance is a triangle-headed arrow. This allows us to see immediately, that both `Tree` and `Flower` have inherited the characteristics of `Plant`. There is no limit to the 'depth' of our inheritance hierarchies, so we can for example, create a further subclass `DeciduousTree` that inherits the characteristics of `Tree` which in turn inherits the characteristics of `Plant`.

When describing inheritance, you will often come across the terms *superclass* and *subclass*. Superclass is the more general of the related classes (`Plant` in our example) and subclass is the more specific (`Flower` and `Tree` in our example). These are also sometimes referred to as the *parent class* and *child class* respectively.

3.1.1 Inheritance in Java

Inheriting classes in Java is a relatively straightforward process. We use the keyword `extends` when we declare our class.

The general form of inheritance is,

```
class subclassName extends superclassName
```

For example,

```
public class Tree extends Plant
{
    // Class definition here.
}
```

Once we have done this, any `public` methods or class variables created in the superclass immediately become available to the subclass. If we had declared any of our methods or variables as `private`, they would not be available to the subclass (but more on this below).

So, let us consider how we might restructure our `Plant` and `Tree` classes to use inheritance:

```
//   *************************************************************
/** Represents any plant. All plants have an age and a height.
  * @author    Jo Wood
  * @version   1.1, 27th August, 2001
  */
//   *************************************************************
```

```
public class Plant
{
    // ------------------ Object variables --------------------

    protected float height;            // Stores height of plant.
    protected int age;                 // Age of plant.
    protected int locationX,locationY; // Location of plant.


    // -------------------- Constructor -----------------------

    /** Initialises the height and age of the plant.
      */
    public Plant()
    {
        age = 0;
        height = 0;
        locationX = 0;
        locationY = 0;
    }

    // --------------------- Methods --------------------------

    /** Doubles the height of the plant and ages it by a year.
      */
    public void grow()
    {
        height = height*2;
        age = age + 1;
    }

    /** Displays details of the plant.
      */
    public void displayDetails()
    {
        System.out.println("Plant is "+age+" years old and " +
                            height+"m tall");
    }
}
```

Our `Plant` class looks pretty similar to the way in which we had previously defined `Tree`. It contains two variables, `age` and `height` since we can assume that *all* plants will have at least these characteristics. We can also assume that all plants can grow at some rate, so we have included an appropriate method inside the class.

You will notice a new keyword `protected` in front of our variable declarations. If a variable is declared `protected` it means that only objects made out of the class itself or any subclasses will know of its existence. `protected` variables can be thought of as being half way between `private` where no other classes know of their existence and `public` where all other classes know of their existence.

```
//  ****************************************************************
/** Represents a Tree with its own height, age and location.
  * @author    Jo Wood
  * @version   1.1, 23rd May, 2001.
```

```
     */
//   ****************************************************************
public class Tree extends Plant
{
     // --------------------- Constructor -----------------------

     /** Creates a new tree and initialises its state.
       */
     public Tree()
     {
         super();         // Call the superclass' constructor
         height = 1;
         age    = 2;
         displayDetails();
     }

     // ---------------------- Methods --------------------------

     /** Grows the tree by 5% and ages it by a year.
       */
     public void grow()
     {
         height = height + (0.05f * height);
         age = age + 1;
     }
}
```

Our `Tree` class is now a little simpler than it was since much of the declaration of variables and methods has been moved to the `Plant` superclass. However, there are still some things that are specific to trees that we have redefined inside this class. We have added a location and initialised `height` and `age` to be more tree-like. We have also slowed down the growth rate of the tree from the 100% per year defined in `Plant` to a more reasonable 5%. This process of redefining a method that has already been created in a superclass is known as *overriding*. Overridden methods (i.e. ones that are defined in a subclass) always take precedence over their definitions in the superclass.

If an appropriate object is created out of the `Tree` class (say `elmTree`), it will be perfectly valid to call the following methods:

```
elmTree.displayDetails();
elmTree.grow();
elmTree.displayDetails();
```

even though one method was defined inside `Plant` and the other inside both `Plant` and `Tree`. This is one of the attractive features of inheritance. When we use a class that inherits another, we need not bother ourselves as to where in the inheritance hierarchy methods are defined.

One word of warning when using inheritance. All public and protected methods of superclasses are automatically made available to a subclass *with the exception of the constructors*. If we wish to call a superclass' constructor, we must do so explicitly using the keyword super. You can see an example of this in the Tree constructor where super() calls Plant's constructor before moving on to the reinitialisation of age and height.

3.2 ABSTRACT METHODS AND INTERFACES

We have seen that object-oriented inheritance allows us to borrow 'for free' the state and behaviour of a previously written class. This is particularly efficient when several classes all inherit from the same superclass. However, we are faced with a problem when designing potential superclasses. We are unlikely to know in advance which other classes may some day inherit the class we are writing. In such cases, we might have trouble coding all of our methods in a suitably flexible way that make them useful to all future subclasses.

To illustrate the point, consider the Plant class coded above. We have created a method grow(), which describes the behaviour that we expect *all* possible subclasses of Plant to exhibit. This is a reasonable assumption and can help us control the type of behaviour of all plant models. The problem arises when we consider exactly *how* potential plants might grow. Trees might increase their height on an annual basis, but a cactus might increase its width, or an algal bloom might increase its surface area daily. We therefore might want to stipulate that all plants must grow, but not *how* that growth behaves. In such cases, we can create special empty methods known as *abstract methods*. These have a name like any other, but contain no code within them. It is up to the inheriting subclass to provide the details of the method's implementation. In Java, we declare an abstract method with the keyword abstract. For example,

```
public abstract void grow();
```

Note that we have a semicolon following the method name and no braces. By declaring one or more methods abstract, the class in which they sit also becomes abstract. This means that we cannot create an object directly out of the class as we have done previously. In order to instantiate a class, we must first subclass it and then create code for at least the abstract methods. For example,

```
public class Cactus extends Plant
{
    private float width;      // Width of cactus in mm.

    public Cactus()
    {
        width = 50;           // Default width of cactus.
    }

    public void grow()
    {
        width += width*0.01f;
    }
}
```

We have effectively overridden the abstract method with our own cactus-specific code. The advantage in this case is that we have not 'wasted' an implementation of `grow()` in the superclass. It also has the conceptual advantage of separating abstract organisation of things from their concrete representation.

Classes with one or two abstract methods are actually quite rare in Java. What is far more common (and powerful) is a class that consists *entirely* of abstract methods. In object-oriented terms, this is known as an *interface* (not to be confused with a Graphical User Interface discussed later). To create an interface in Java we (not surprisingly) use the keyword `interface` in place of `class` when declaring the class.

The general structure of an interface declaration is

```
public interface InterfaceName
{
    public abstract methodName1();
    public abstract methodName2();

    // Any other methods declared in a similar way.
}
```

where `InterfaceName` is the name of the interface to create using the same naming conventions as any other class.

Creating a class that uses an interface is similar to inheriting a class, with two important differences. First, we use the Java keyword `implements` rather than `extends`. By implementing an interface, we are committing ourselves to create code for all the abstract methods within the interface. Second, unlike inheritance, we can implement as many interfaces as we wish. So, the following will be perfectly acceptable in Java:

```
public class ParkKeeper extends Person
                        implements Mobile, Employed
{
    // Class definition here.
}
```

In this example, `Mobile` and `Employed` are two interfaces declared elsewhere that identify certain types of behaviour without specifying the details of what that behaviour entails. For example, the `Mobile` interface might include the method `move()` which changes the spatial location of the object implementing it. `Employed` might contain an accessor method `getSalary()` that guarantees that any employed object can tell us about its annual salary.

You may be wondering why go to all the trouble to add empty methods to a class, if we have to add our own method code anyway. The answer is suggested by the name *interface*, since we can guarantee that any class that implements an interface will contain the methods defined within it. This allows us to predict to a certain extent, what type of behaviour our new class will have.

The creation of abstract data types (ADTs) and interfaces is an entire branch of computer science in itself and we will not dwell on it here. We will however show how interface design and implementation can be used in the case study at the end of this chapter.

3.3 PASSING MESSAGES

So far, we have related our classes to each other by using *composition* and *inheritance*. This allows classes to be linked at the design stage and is likely to reflect the way in which entities are related in our object-oriented model.

A third and important part of object-oriented modelling allows us to relate *objects* together by providing a way for objects to send information to one another. This process is known as *message passing* and involves creating methods in Java to handle the flow of information between objects and is closely related to the object-oriented process of *encapsulation*.

When we design our classes in Java, one of the goals we aim for, is to make each class as independent of all other classes as possible while still functioning effectively. For example, once we have instantiated a `tree` object (from the class `Tree`) it should go about its business of growing, shedding its leaves by itself. Our object-oriented model of the park is not concerned with the precise details of how this happens, just as long as we can find out how tall it is at any given time and if we need to sweep up any fallen leaves.

The process of hiding the internal workings of an object from the outside world is known as *encapsulation*. This is a very useful idea because it can make the process of programming much simpler if we are spared the details of things we are not concerned about. In Java, there are various ways in which we can hide the details of the workings of a class, but one of the main ways we can do this is to make sure that all variables declared outside a method (e.g. `age` and `height` in our plant/tree example) are declared `private`.

So, if we hide all the workings of a class, how can we find out information about it? The answer is that objects can send messages to each other. For example, we might send a message to a tree object and get another in return (see Figure 3.2). It should not matter *how* the `oakTree` object finds out how tall it is, just as long as it reports its height back to us.

We can send information both to an object (the messages on the left in Figure 3.2) and get an object to send back new information (the messages on the right in Figure 3.2).

Developing Classes and Objects

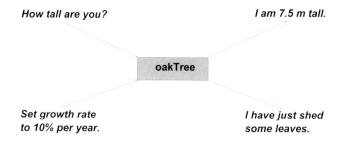

Figure 3.2 Passing messages to and from an object.

3.3.1 Sending Messages in Java

We now have a way to define class descriptions and create objects that are specific instances of those classes. We have managed to encapsulate the state of each object by declaring all the variables inside it as either `private` or `protected`. This means that we cannot read or modify any of the variables in `Tree` from any object that itself is not of type `Tree`.

Suppose that we wish our `Park` class to be able to modify the location of any `Tree` object. How could we do this? The answer is to send a message to the `Tree` object telling it to change its location. In fact, it would also be quite useful to allow the `Tree` object to send a message saying what its current location is.

Because both of these messages involve some action (changing tree location and reporting tree location), we can deal with these messages by creating some new methods inside the `Tree` class:

```java
// ------------------- Mutator Methods ----------------------

/** Sets a new location for the tree.
  * @param newX x-coordinate of the new tree location.
  * @param newY y-coordinate of the new tree location.
  */
public void setLocation(int newX, int newY)
{
    locationX = newX;
    locationY = newY;
}

// ------------------- Accessor Methods ----------------------

/** Reports the current x-location of the tree.
  * @return x-location of the tree.
  */
public int getLocationX()
{
    return locationX;
}
```

```
/** Reports the current y-location of the tree.
  * @return y-location of the tree.
  */
public int getLocationY()
{
    return locationY;
}
```

The method `setLocation()` allows us to change the values of the two variables `locationX` and `locationY`. We can do this by declaring two *parameters* inside the brackets following the method name. Parameters can be used to hold any incoming messages to an object and are declared just as we would any other variable. Methods such as this one which change the state of a class in some way are known as *mutator methods* (see Figure 3.3).

The methods `getLocationX()` and `getLocationY()` do not change the state of the object, but rather report aspects of its current state. These are known as *accessor methods* (see Figure 3.4). These are declared in a similar way to any other methods except that after the word `public` we state the *type* of message that will be returned by the method. In our example above, these will be whole number integers. Additionally, we include within the method, the keyword `return` followed by the message we wish to be reported by the method. `return` only allows a single value to be reported, so we need two methods to report both the x and y coordinates of the location.

This process explains why we have been previously using the term `void` in our method declarations. `void` simply means that no message is to be returned by the method.

Method Signatures and Local Variables

A method inside a class can always be identified by its name and the type of incoming and outgoing messages it receives or sends. It is also possible to define more than one method with the same name inside a class, providing each has a unique *signature*.

A signature is defined by the type and number of incoming parameters received by a method. For example, the signature of the following method is (`int`, `int`)

```
public void setLocation(int row, int col)
```

and is distinct from the following which has a signature of (`float`, `float`)

```
public void setLocation(float easting, float northing)
```

However, if a class also contained the method

```
public void setLocation(float x, float y)
```

Developing Classes and Objects

> the Java compiler would complain that two methods had been defined with the same signature (`float, float`). Note that the names of the incoming parameters (as distinct from their type) have no bearing on the uniqueness of the signature. This is because names given to incoming messages are known as *local variables* and can only be used within the method into which they are placed.

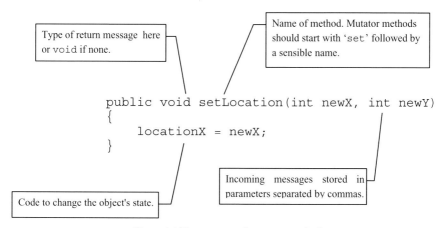

Figure 3.3 The structure of a mutator method.

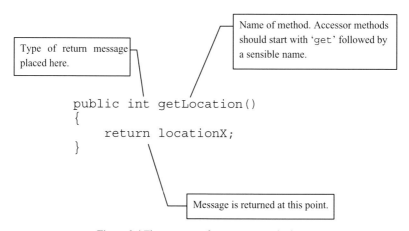

Figure 3.4 The structure of an accessor method.

Once we have created our methods, we can call them just as we would any other method. When we wish to pass a message into an object, we simply call the relevant method and substitute values for the parameters declared in the method. To retrieve a message from an object, we again call the relevant method, and treat it as if it was a variable capable of storing a value of some kind.

So, for example, we could create an oak tree object, place it at a given location and display its position as follows:

```
Tree oakTree = new Tree();

int xCoord,yCoord;                  // Location.
oakTree.setLocation(356000,408650);
xCoord = oakTree.getLocationX();
yCoord = oakTree.getLocationY();

System.out.println("Tree is located at " +
                    xCoord + "," + yCoord);
```

Note that the names we give to both the incoming and outgoing message variables are completely independent of the ones defined with the method.

You may be asking yourself, why go to the bother of creating accessor and mutator methods just to read and change the value of a private variable? Will it not be better to make the variable public in the first place so that other objects can manipulate its value directly?

The reason that good object-oriented design should use accessor and mutator methods in combination with private variables is perhaps best illustrated with an analogy. Imagine you wish to find out the National Insurance details stored about yourself by the government (name, address, National Insurance payments etc.). You can do this in two ways. First, you may visit the relevant government building, go to the office where your records are stored, go straight to the filing cabinet with your file in it and examine its contents. You may even decide to change the contents (for example by decreasing the amount of National Insurance contributions you are required to make).

Disregarding the fact that such details are unlikely to be stored in filing cabinets, this approach has several obvious problems. First, you may not know where to look to find your file. Second, you may not have the privilege to change the contents of your file. If an employer or tax office need to use the details on your file, they may not know to update details stored elsewhere that have been changed by you.

Far better would be to make a request of someone working there to give you the information on your file, and perhaps ask them to update some of its details. This latter approach is analogous to calling an accessor and mutator method whereby access to private data is controlled by routing all requests and updates through known methods (government office workers). Problems of data integrity, and security are then minimised.

Developing Classes and Objects 55

3.4 USING GRAPHICAL CLASSES IN JAVA

One of the first large class hierarchies you are likely to come across when creating Java programs is the package of classes that make up Java's graphical user interface components. These classes allow you to create window-based graphical 'front ends' to your programs. We will revisit Java's graphics functionality throughout this book, but as an introduction we will consider the scope and organisation of graphical components available to us as programmers. In doing so, it will demonstrate the importance and power of object-oriented modelling in representing complex groups of entities.

3.4.1 Swing and the AWT

When Java was first released in 1995, it was one of the few languages that contained its own graphics functionality as part of the core language available for all supported platforms. The classes that allowed graphical features to be created were collectively known as the *Abstract Windowing Toolkit* or AWT. This toolkit was designed to implement only the graphics functionality that was likely to be available on all platforms. Consequently, it represents a 'lowest common denominator' of the windowing environments upon which Java runs (Windows, Unix and MacOS being the most common). Figure 3.5 shows a sample of some of these AWT components.

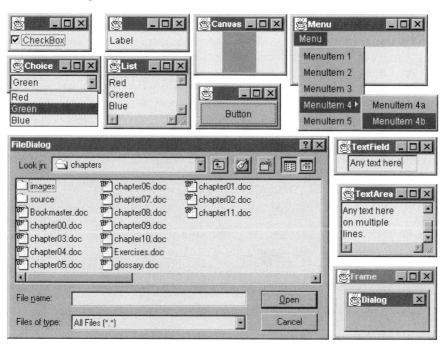

Figure 3.5 Sample AWT components.

The look and feel of the components shown in Figure 3.5 will be familiar to anyone who uses a Windows operating system on a PC. They are the components that are created by the Windows environment complete with any user-preferences such as font settings, window colours etc. The same components would have a 'Mac look and feel' if rendered in the MacOS environment, or 'K-desktop look and feel' on a Linux platform (see *Heavyweight and Lightweight Components* below). This has the immediate advantage of allowing a familiar user interface to be created in Java, but tends to limit the control the programmer has over the way the components are rendered.

The AWT uses inheritance extensively because many of the graphical classes that it contains have much in common with each other (for example, the way in which they can be moved, resized and repainted). Figure 3.6 shows a class diagram representing most of the AWT graphical components. Most of these components are inherited from the abstract class `Component`. This is an example of inheritance generalisation where functionality common to graphical components is stored in a single class from which all others are derived. You will notice that the AWT treats menus in a slightly different way as all menu classes are inherited from `MenuComponent`.

The AWT is still used in Java, but it does suffer from one major drawback. All AWT components are drawn by the native operating system (i.e. Java passes responsibility for drawing these components to Windows on a PC or the window manager on a Unix platform). While this has the advantage of speed and user familiarity (see *Heavyweight and Lightweight Components*) it does result in a rather limited graphical functionality. Additionally, the object-oriented design of the AWT is rather poor in places (for example, placing the menu components in their own hierarchy) making it somewhat inflexible and difficult to understand.

Partly for these reasons, Java provides an additional set of classes known as the *Swing Toolkit*, which provides much richer and better designed graphical functionality. Unlike the AWT, all Swing components are drawn and handled by Java, giving much more control to the programmer. Simplified class diagrams of some of the Swing components are shown in Figures 3.7 and 3.8. Swing components can be immediately distinguished from their AWT counterparts as they tend to be prefixed with a 'J'. Notice also how all Swing components are inherited from one of four AWT 'containers', either `Dialog`, `Window`, `Container` or `Applet`.

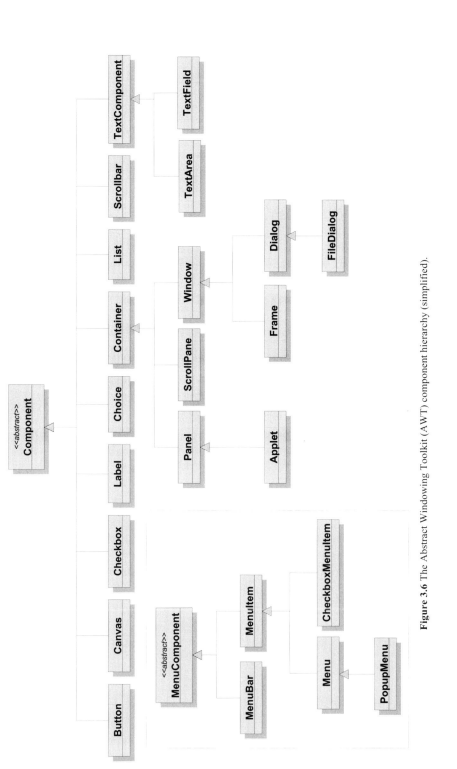

Figure 3.6 The Abstract Windowing Toolkit (AWT) component hierarchy (simplified).

Heavyweight and Lightweight Components.

There are two sets graphical components in Java, known as *heavyweight components* and *lightweight components*. Heavyweight components such as those in the AWT are so-called because they use the native operating system to create the graphics. This means that the menus, buttons etc. will look like any other on the same system (same colours, fonts etc.). The advantage of heavyweight components is that they tend to be drawn quite fast and are likely to be familiar to the user.

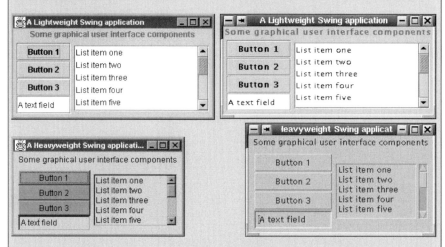

Lightweight components are those that are drawn directly by Java. These do not rely on calls to the native operating system and will look very similar on all platforms. The most common collection of lightweight components in Java is the so-called 'Swing' package.

The choice of whether to program using AWT or Swing will depend on a number of factors. Perhaps, the most influential is whether or not you are intending to create an Applet embedded in a web browser (see Chapter 10). Most browsers will display AWT components, but will not display Swing graphics unless a *Java Plugin* is loaded. Therefore, small Applets designed for a large client base should probably be coded using the AWT.

If you are intending to create a sophisticated graphical user interface, Swing is probably the more sensible toolkit to use. Despite its wealth of available classes making this seem like the more complicated choice, its good object-oriented design makes constructing sophisticated interfaces an easier task.

Heavyweight and lightweight components should not be mixed in the same application as this not only creates a confusing GUI, it can lead to problems when they are drawn on screen together.

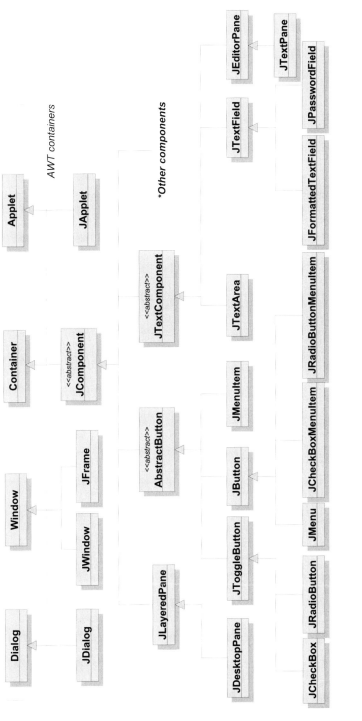

Figure 3.7 The Swing component hierarchy (simplified). *See Figure 3.8 for a list of other components.

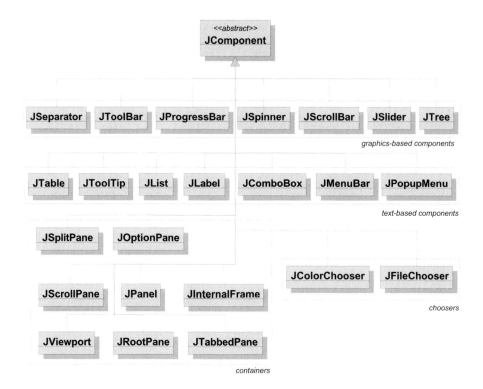

Figure 3.8 Swing components inherited directly from JComponent

3.4.2 Creating Simple Graphics

Let us consider how we might make some sense of these classes when constructing our own graphical programs. As an example, consider the process of creating a simple closable window containing a simple message.

```
import javax.swing.*;        // For Swing components
import java.awt.*;           // For AWT components.

// ***************************************************************
/** Creates a simple Swing-based window
 *  @author   Jo Wood.
 *  @version  1.2, 3rd September, 2001
 */
// ***************************************************************

public class SimpleWindow extends JFrame
{
```

Developing Classes and Objects

```
    // -------------------- Object Variables --------------------
    private Container contentPane;    // This is the container into
                                      // which components are added.

    // --------------------- Constructor ---------------------

    /** Creates a simple window with an equally simple label
     * inside.
     */
    public SimpleWindow()
    {
        // Create the window with a title.
        super("A simple window");
        setDefaultCloseOperation(WindowConstants.DISPOSE_ON_CLOSE);
        contentPane = getContentPane();

        // Create a label to add to the window.
        JLabel label = new JLabel("This is a very simple window");
        contentPane.add(label);

        // Get Java to size the window and make it visible.
        pack();
        setVisible(true);
    }
}
```

In order to 'run' this program, we need to create a `main()` method. This can either be within `SimpleWindow`, or in its own class.

```
// ****************************************************************
/** Creates an object out of the SimpleWindow class. Displays a
  * very simple window with an equally simple message.
  * @author Jo Wood.
  * @version 1.2, 25th August, 2001
  **/
// ****************************************************************
public class RunSimpleWindow
{
    /** Creates a simple window.
      */
    public static void main(String args[])
    {
        new SimpleWindow();
    }
}
```

When we run `RunSimpleWindow`, the output shown in Figure 3.9 is produced.

Figure 3.9 Output from `SimpleWindow.java`.

Notice that for any graphic programming to work, we must include the lines `import java.awt.*` and `import javax.swing.*`. We will examine importing packages in detail in Chapter 6, but for the moment these two lines can be thought of as issuing the instruction 'add graphics classes to this one'. Note also that even though we are using Swing components, we need to import both the AWT and Swing packages.

This class, like most graphics programs involves inheriting one of the existing top-level container classes. `SimpleWindow` inherits the `JFrame` class, which is the main class for creating top-level windows with Swing. The constructor uses the Java keyword `super`, which simply calls the superclass' (`JFrame`) constructor as we saw in Section 3.1.1. We know that the window initialisation code will be dealt with somewhere in the class's hierarchy. In fact, you may have noticed from the class diagrams in Figure 3.7 that `JFrame` inherits the `Window` class from the AWT and it is here that the window's initialisation is carried out. The only extra initialisation we do is explicitly tell the window that it should close itself down if closed by the user (usually by clicking the cross in the corner of the window). We do this by calling the method `setDefaultCloseOperation()` that is part of the `Frame` class.

The technique for building graphical user interfaces in Java involves creating graphical components and getting Java to place these inside one of the 'container' classes such as `JFrame`. This is done by extracting what is called a *content pane* from the container, and adding components to it. In our example, we have stored the content pane in its own object variable (`contentPane`). Content panes (which are of type `Container`) have their own method called `add()` which allows any other graphical components to be placed inside them. The example above creates a `JLabel` object and adds it to `contentPane`.

We must get Java to size the window so that it is just large enough to hold the components we have added to it. The `pack()` method does this and saves us the effort of working out the window dimensions ourselves. Finally, we make the window visible by calling the `setVisible(true)` method. By default, all containers will be invisible until they are made visible with this method. This allows us to create the window with all its contents before anything is seen by the user.

In the next chapter, we will develop the idea of laying out components inside containers using what are known as *layout managers*.

3.5 CASE STUDY: MODELLING ANTS IN THE GARDEN

We shall develop some of the ideas covered so far by designing and building a set of Java classes to model the behaviour of ants living in a garden. You will recall from the first chapter that we used this example to illustrate the advantages of the object-oriented modelling approach. Here, we shall elaborate on the design procedure and implement the design with a set of simple Java classes. We will

Developing Classes and Objects 63

revisit this case study throughout the book until we have a fully functioning model that allows us to examine ant behaviour graphically.

3.5.1 Programming Objectives

As with all well thought out programming processes, the first task is to identify our programming objectives explicitly. In our case, we must write a program to '*model ants in a garden*'. This is obviously not specific enough to implement as it stands, so we will identify a series of more implementable programming objectives:

- to create a graphical environment that simulates the movement of ants within a confined space;

- to model the moving and feeding behaviour of each ant in a reasonably realistic manner; and

- to model a central 'nest' from which ants emerge and explore the surrounding surface.

Our first attempt to meet the programming objectives should be to construct a simple object-oriented model of the environment. Time spent designing the model at this stage is likely to save much programming time later on, so it is worth spending some effort in creating an appropriate model. Do our programming objectives suggest any obvious classes of objects to construct?

Three classes might initially spring to mind, namely classes representing *ants*, *nests* and *gardens*:

Ant	Nest	Garden

State:
  location
  foodLevel
	State:
	  location
	  population
	  foodStore
		State:
		  location
		  boundary
		  contents

Behaviour:
  eats
  moves
  evolves
	Behaviour:
	  evolves
		Behaviour
		  evolves

Before we go any further, we should consider whether our classes have anything in common with each other. If they do, we should contemplate extracting their common elements and placing them in separate classes. This has the advantage of reducing repetition of coding and increasing the *reusability* of our classes.

Two obvious areas of repetition should be apparent from our initial class design. All classes have some form of spatial representation (location or boundary), and all evolve over time. It makes sense to encapsulate both of these characteristics within their own classes. The process has further reuse benefits since we can anticipate that both a spatial object and dynamic object can be used in a variety of other contexts. We shall give our spatial class the ability to store a location and a boundary as well as perform a simple spatial comparison (for example, to find out if one spatial object is *within* another, or *overlapping* with another etc.). For the moment, we shall just give our dynamic class the ability to 'evolve' in some way over time.

SpatialObject

State:
  location
  boundary

Behaviour: compare

<<interface>> Dynamic
State:
Behaviour: evolve

3.5.2 Relating Classes

Now that we have defined five simple classes, our next task is to identify any links between them. We have already established that `Ant`, `Nest` and `Garden` are all related to `SpatialObject` and `Dynamic` in some way, but what is the nature of that relation? As we saw earlier in the chapter, whenever we wish to relate classes together, we have a choice of three options: (1) *inheritance*; (2) *composition*; or (3) *implementation*.

Composition is used when one object uses the functionality of another, but does not wish to make its interface public. Inheritance and implementation are used when the public interface of the existing class should remain public within the new one. The difference between the two is that implementation of a public interface involves providing code to define the implemented behaviour, whereas with inheritance, the state and behaviour of the inherited class will have already been coded.

In our example, we will let each of our first three classes inherit the `SpatialObject` class since the '*is a*' relationship seems most natural for our model. Inheritance is particularly appropriate when the inherited behaviour is likely to be very similar in all classes.

We will let the same three classes implement the `Dynamic` interface because, although all will *evolve* in some way, the nature of that dynamic change is likely to be different for each class. Implementation tends to be used when we wish several

Developing Classes and Objects

classes to have some named behaviour in common, but the precise nature of that behaviour varies between classes.

Finally, we will use composition to define the relationship between the `Garden`, `Nest` and `Ant`. Composition tends to be used when we wish to model one class containing another in some way. In our example, that containment is a spatial one since the garden '*has a*' nest and the nest '*has a*' [some] ant[s].

We can now represent all our class definitions in a class relationship diagram. Figure 3.10 shows the representation of our class structure so far. Notice that the convention of representing an interface using a rounded rectangle.

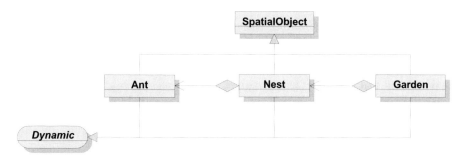

Figure 3.10 Class diagram showing simple design for ants in the garden model.

3.5.3 Maximising Class Reuse

Should we now start coding our classes in Java? Not quite, because we can anticipate that our final model will be more complex than the one described so far. If it does become more complex, we would like to reuse as much of our existing class structure as possible, so we should look for further ways of economising on our programming effort. We know that our ants will have to eat and move about the garden, but these behaviours are ones that may be shared by other animals in the garden. We may therefore define a new class `Animal` that does things that all animals do, and let our `Ant` class define ant-specific activities.

For similar reasons, we can improve the efficiency of our spatial modelling by clarifying the difference between a `SpatialObject`'s state and behaviour. A spatial object might be a single 2D point, a discontinuous 3D volume, or something more complex still. Rather than trying to define these possibilities at the outset of our design process, we can allow `SpatialObject` to be *composed* of a new class `Footprint` that deals with the necessary complexities of the geometry. By separating the geometry from the behaviour, we allow other classes to store and exchange spatial geometry without the additional overhead of the spatial comparison behaviour of `SpatialObject`.

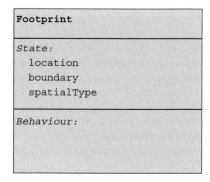

Finally, we will make one further change that should promote reusability of our classes in the future. We can identify the behaviours that we think might be adopted by more than one class, but implemented in different ways for each. As with our Dynamic interface, we declare (but do not implement) such behaviours in interfaces that have no state and may be implemented by a range of other classes. In our example, it makes sense to add one further interface to describe feeding behaviour. We know that animals and plants all feed, but they are likely to do so in different ways. Initially, this interface will be implemented by our Animal class.

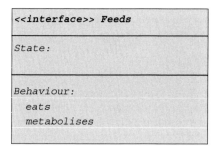

Our initial class design containing the two interfaces, which is shown in Figure 3.11, is now ready for coding.

Developing Classes and Objects

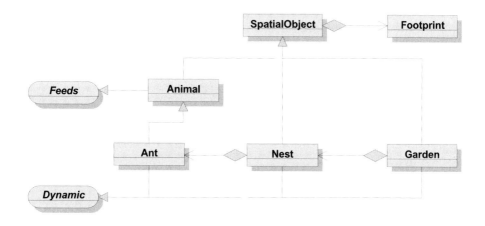

Figure 3.11 Ants in the garden class design ready for coding in Java.

3.5.4 Coding the Class Design

Our two interface designs are straightforwardly converted into Java. The `Dynamic` interface is very simple, containing the single method `evolve()`. This is the method that we guarantee all dynamic objects contain, and may be called whenever we wish the object that implements it to evolve in some way.

```
//   ********************************************************
/** Interface for things that evolve through time.
  * @author Jo Wood
  * @version 1.1, 17th January, 2000.
  */
//   ********************************************************
public interface Dynamic
{
    /** Defines the changes that occur to the object over time.
      * This method should be called whenever time advances by
      * 1 unit.
      */
    public abstract void evolve();
}
```

The `Feeds` interface ensures that all objects that implement it define some eating and metabolising behaviour.

```
//   ********************************************************
/** Interface for things that can eat.
  * @author Jo Wood
  * @version 1.1, 17th January, 2000.
  */
//   ********************************************************
```

```
public interface Feeds
{
    /** Eat a given amount of food.
      * @param foodUnits Amount of food to eat.
      */
    public abstract void eat(int foodUnits);

    /** Metabolise a given amount of food.
      * @param foodUnits Amount of food to metabolise.
      */
    public abstract void metabolise(int foodUnits);
}
```

To see how these two interfaces are implemented, consider simple representations of the `Animal` and `Ant` classes. Initially, the design of both classes will be quite similar, but by keeping them separate we keep the design open for future program development. Their simple implementations are given below:

```
//      ********************************************************
/** Class for defining an animal. All animals are spatial
  * objects and can live, die and feed in some way.
  * @author Jo Wood
  * @version 1.2, 23rd August, 2001.
  */
//      ********************************************************
public class Animal extends SpatialObject implements Feeds
{
    // ------------------ Object variables ------------------

    private int foodLevel;    // Food level of animal.
    private boolean alive;    // Is animal dead or alive?

    // --------------------- Constructor --------------------

    /** Creates animal with given initial food level and footprint.
      * @param foodLevel Initial food level of animal.
      * @param footprint Animal's initial footprint.
      */
    public Animal(int foodLevel, Footprint footprint)
    {
        super(footprint);
        this.foodLevel = foodLevel;
        alive = true;
    }


    // ---------------------- Methods -----------------------

    /** Let the animal eat some food and thus increase its food
      * level by the given value.
      * @param foodUnits Number of food units to eat.
      */
    public void eat(int foodUnits)
    {
        foodLevel += foodUnits;
    }
```

```
    /** Let the animal metabolise some food and thus decrease its
      * food level by the given value.
      * @param foodUnits Number of food units to metabolise.
      */
    public void metabolise(int foodUnits)
    {
        foodLevel -= foodUnits;
    }


    // ---------------- Accessor/mutator methods -------------

    /** Report the current food level of this animal.
      * @param return Current food level.
      */
    public int getFoodLevel()
    {
        return foodLevel;
    }

    /** Set the current food level.
      * @param int foodLevel New food level for object.
      */
    public void setFoodLevel(int foodLevel)
    {
        this.foodLevel = foodLevel;
    }

    /** Determines if animal is alive or not.
      * @return True if animal is alive.
      */
    public boolean isAlive()
    {
        return alive;
    }

    /** Kills the animal (or raises from the dead).
      * @param alive Animal is dead if false, alive if true.
      */
    public void setAlive(boolean alive)
    {
        this.alive = alive;
    }
}
```

This class implements simple eating and metabolising activity by incrementing an internal food level every time the animal eats and decrementing it every time it metabolises. If we later subclass Animal we may substitute more complex activity, but this should be acceptable as a default behaviour.

The Ant class does little more than the more general Animal class at the moment. The only addition is a simple implementation of the evolve() method that all Dynamic classes must have. While we have not decided how an ant will move yet, we can assume that it can only move and metabolise if it is alive, and that merely to exist involves metabolising the ant's own food supply.

```
//   ***********************************************************
/** Class for defining Ants.
  * @author Jo Wood
  * @version 1.2, 23rd August, 2001.
  */
//   ***********************************************************

public class Ant extends Animal implements Dynamic
{
    // -------------------- Constructor ---------------------

    /** Creates ant with given initial food level and footprint.
      * @param foodLevel Initial food level of the ant.
      * @param footprint Initial spatial footprint of the ant.
      */
    public Ant(int foodLevel, Footprint footprint)
    {
        super(foodLevel,footprint);
    }

    // ------------------ Implemented methods ----------------

    /** Let the ant go about its business for one time unit.
      */
    public void evolve()
    {
        metabolise(1);              // Use up 1 food unit.

        // Add ant movement code here.
    }
}
```

The Nest and Garden classes act as spatial 'containers' doing little other than being composed of other dynamic spatial classes. The Garden class contains an ant's Nest and the Nest class contains a collection of one Ant. We will see later how we can add more ants to the collection.

As each class uses the Dynamic interface, we must implement the evolve() method in each, as we did for the Ant class.

```
//   ***********************************************************
/** Class for defining an ants' nest.
  * @author Jo Wood
  * @version 1.2, 23rd August, 2001.
  */
//   ***********************************************************

public class Nest extends SpatialObject implements Dynamic
{
    // ------------------ Object variables ------------------

    private Ant ants;       // A colony of 1 ant.


    // -------------------- Constructor ---------------------
```

```java
    /** Creates a nest with a given spatial footprint.
      * @param footprint Spatial footprint of the nest.
      */
    public Nest(Footprint footprint)
    {
        super(footprint);

        // Add 1 ant with 100 food units at the nest location.
        ants = new Ant(1000, new Footprint(footprint.getXOrigin(),
                                           footprint.getYOrigin(),
                                           8,5));
    }

    // ----------------- Implemented Methods -----------------

    /** Let the nest evolve for one time unit.
      */
    public void evolve()
    {
       ants.evolve();
    }
}
```

```java
// ***********************************************************
/** Class for defining a garden containing an ants' nest.
  * @author Jo Wood
  * @version 1.2, 23rd August, 2001.
  */
// ***********************************************************

public class Garden extends SpatialObject implements Dynamic
{
    // ----------------- Object variables -------------------

    private Nest antsNest;    // A colony of ants.

    // ------------------- Constructor ----------------------

    /** Creates a default garden with a given spatial footprint.
      * @param footprint Spatial footprint of garden.
      */
    public Garden(Footprint footprint)
    {
        super(footprint);
        antsNest = new Nest(new Footprint(100,80));
    }

    // ----------------- Implemented methods -----------------

    /** Let the garden grow for one time unit.
      */
    public void evolve()
    {
        antsNest.evolve();
    }
}
```

We will defer discussion on how we code the spatial classes until we have covered some further Java concepts. But what we have so far provides us with an extensible and reusable set of Java classes upon which we can develop our ant model.

3.6 SUMMARY

In this chapter, we have identified the difference between general classes of things and specific instances of those classes called objects. We have continued to develop the process of object-oriented modelling by discussing three new important concepts, namely *inheritance*, *encapsulation* and *interfaces*. These allow classes in an object-oriented model to be linked together.

Inheritance has allowed us to organise our classes in a more efficient manner by identifying the aspects that different classes have in common and representing them in their own class. Classes linked using inheritance are sometimes said to have an 'is a' relationship – for example, 'tree is a plant'.

The implementation of an interface whereby a class containing empty abstract methods is inherited, is sometimes referred to as an 'acts as a' relationship. Interfaces are useful in object-oriented design as they allow us to specify what a class does without having to give details about how it is done.

Encapsulation is useful in hiding the detailed workings of a class and only 'exposing' those bits of it necessary for communication with other classes and objects. We have seen that to build a manageable object-oriented model, we try to keep classes as self-contained and as private as possible. But, we have also seen that to make the model work effectively, we must get objects to communicate with one another. This can be done by passing messages both to and from public methods with a class.

These important object-oriented ideas were illustrated by introducing Java's graphics classes contained in the AWT and Swing packages. Finally, the process of constructing an object-oriented model was demonstrated with the design of a simple ant simulation model.

By the end of this chapter and associated exercises on the web, you should be able to

- understand the difference between a class and an object;
- recognise that difference when reading Java code;
- understand what a constructor is and what type of code should be placed inside it;
- create a simple object-oriented model that uses inheritance to define some of the classes;
- create public accessor and mutator methods to read and change private variables; and
- create a very simple graphical class using the Swing toolkit.

CHAPTER FOUR

Controlling Program Movement

One of the characteristics of languages such as Java that contain procedural elements, is that they consist of an ordered series instructions to the computer that describe how to achieve something. It has been possible to predict the order in which these instructions will be executed by reading 'down the page'. In the examples we have seen so far, each instruction has been executed once and once only.

However, programming tasks frequently involve repeating certain instructions many times over. This chapter considers how we can implement repetitive program instructions through the use of loops. This process is illustrated with the development of a simple 'raster' model of spatial information.

4.1 PROCEDURAL AND OBJECT-ORIENTED DESIGN REVISITED

Procedural programming languages can be contrasted with another approach known as *declarative programming*. Declarative languages such as SQL, Prolog or Lisp, do not consist of instructions that describe how to do something, but rather a series of 'questions' that when combined with a data store of some kind, can be used to extract useful information.

In fact, object-oriented programming involves aspects of both procedural and declarative approaches. We have seen that our object-oriented models are less concerned with *how* something is achieved than with declaring *what* can be achieved. Yet at some level, as we shall see in this chapter, we are also forced to describe line by line, the detailed procedural instructions necessary for achieving our programming goal.

4.2 APPLYING OPERATORS TO VARIABLES

So far, we have been able to declare and use variables that hold integer and decimal numbers. We have seen how we can apply simple operators to those variables such as adding them together. An operator is simply something that modifies a number or variable. In this section, we will consider two types of operator: (1) *arithmetic operators* that are used with numbers of some kind; and (2) *logical operators* that are used on logical or Boolean expressions.

4.2.1 Simple Arithmetic Operators

We have already used these expressions in our programs

+ - * /   (plus, minus, times, divide)

In addition, Java can use one more arithmetic operator, known as the *modulus* operator. The modulus of a number is its remainder after division by another, and is given the symbol %.

For example,

17%5 is 2 (because 17/5 is 3 with a remainder of 2).

This operator only really makes sense when applied to integer expressions. It has a variety of uses, such as determining if a number is odd or even (%2), performing a task on every *n*th number (%n), or extracting the last digit of a number (%10).

4.2.2 Applying Operators to Variables

In the examples we have seen so far, we have applied mathematical operators to variables in the following way:

average = (acidityOld + acidityNew) / 2.0;

where the expression on the right of the = symbol is evaluated and the result is placed in the variable on the left of the =.

We can also combine the same variable on both sides of the equality

total = total + 1; (whatever the variable total contains, make it one bigger.)

or perhaps in a more complex way:

popln = popln + ((growth*popln)*(1-popln));

> The expression above is an example of a simple ecological growth model that can be used to represent the dynamic population change in a community. It represents a 'feedback' loop where the population at any one moment in time is a function of the previous population. Such feedback expressions can be used to model complex *dynamical systems*.

Controlling Program Movement

In Java there exists a commonly used shorthand for several of these expressions as shown in Table 4.1.

Table 4.1 Abbreviated arithmetic operators

Longhand	Shorthand	Description
`total = total + 1;`	`total++`	increase total by 1
`total = total + 7;`	`total += 7`	increase total by 7
`total = total - 1;`	`total--`	decrease total by 1
`total = total - 12;`	`total -= 12`	decrease total by 12
`total = total * 4;`	`total *= 4`	multiply total by 4
`total = total / 2;`	`total /= 2`	halve the value of total

> **Equality and Assignment Operations**
>
> Those with a mathematical training might feel a little uncomfortable with expressions such as those seen in Section 4.4.2. They do not express equalities in the mathematical sense, but rather they symbolise a computational *operation* (take whatever is on the right of the = and place it in the variable on the left). When describing an *algorithm* (detailed description of the procedure followed by a program), such an operation is sometimes symbolised using a left-pointing arrow. For example,
>
> `total ← total + 1`
>
> When we wish to compare two values to see if they are equal (equality operator), Java uses the double equals symbol ==. This is explored in Section 4.2.3 when considering logical operators.

If you really want, these operators can be used in more complex statements.

For example,

`answer = downCount-- + upCount++;`

Note that the ++ and -- operators increment/decrement the variable *after* evaluation of the whole expression. To force increment/decrement *before* evaluation, the operator is placed before the variable.

For example,

`answer = --downCount + ++upCount;`

The expression above is an example of efficient but poor programming practice. It can be confusing and complex to debug. The best advice is to avoid using the shorthand operators in complex expressions, although simple operations like incrementing a single variable by 1 often benefit from the shorthand version.

Never, ever, include a line of code that looks like this (which works by the way):

```
i-=-i-- - --i;
```

> *What would i be equal to if it was initially set to 27?*
>
> *Unless you are particularly gifted, working out the answer to this question will take you some time. The point of the question is to encourage you to avoid expressions like this that might be straightforward for the computer, but very hard for us to understand.*
>
> *(The answer is 79 by the way.)*

4.2.3 Simple Logical Operators

Java contains another type of variable specifically for evaluating logical expressions. A logical expression is one that can have one of two possible outcomes, either `true` or `false`. Expressions that are either true or false are known as *Boolean*.

> **George Boole**
>
> The term Boolean is named after the nineteenth century mathematician George Boole and his works *The Mathematical Analysis of Logic* (1847) and *An Investigation of the Laws of Thought* (1859).
>
> See 'References and Further Reading' at the end of this book for more details.

In Java, you simply use the word `boolean` with a variable as you would for `int` or `float`.

For example,

```
boolean floodRisk = true;
boolean groundFloor = false;
```

The result of applying logical operators to expressions is always itself either `true` or `false`. Examples of logical operators are shown in Table 4.2. These operators will be covered in more detail when we consider looping structures in Section 4.4.

Table 4.2 Logical operators

Symbol	Operator	Example
&&	AND	floodRisk = (inFloodPlane && stormEvent);
||	OR	floodWarn = (inFloodPlane || stormEvent);
==	compare	startDefence = (waterLevel == 2000);
!=	not equal	compensationClaim = (floodDamage != 0);
>	greater than	compensationGranted = (floodDamage > 1000);
<	less than	
>=	greater than or equal to	flood = (waterLevel >= 3200)
<=	less than or equal to	

4.3 RULES OF PRECEDENCE AND PRECISION

4.3.1 Precedence

When we look at some mathematical expression, we tend to evaluate it in parts, usually from left to right. Spaces sometimes give an indication of any variation from this 'rule'.

For example, we might assume the following expression would be equal to 49,

```
3+4 x 3+4
```

However, the Java compiler does not notice spaces in the same way. Further, it ranks all operations in its own order of precedence. For example, * is evaluated before +, so that

```
3+4 * 3+4 actually equals 3 + (4*3) + 4 = 19
```

Sometimes this makes sense, sometimes it does not. As there are so many operators available to the programmer, remembering the order of them is difficult. Therefore, it is better to remember only one rule: *when in doubt, use brackets*.

For example,

```
answer = (3+4) * (3+4);
```

Brackets are always evaluated first, from the 'inside-out'. You should also make things clearer by using spaces to emphasise the order of calculation. As you become more familiar with using operators, you may be tempted to drop the use of brackets. While such expressions without brackets may make sense to *you*, others may find them difficult to deconstruct.

4.3.2 Precision

> *After the following line of code in Java, what is the value of* `answer`?
>
> `float answer = (1/2) + (1/2);`

The answer is not 1 as we might expect, but 0. Because all the numbers to be operated upon (known as arguments) are integers, Java calculates the answer as an integer, even though it is ultimately stored as a floating point number. Consequently, (1/2) is rounded down to 0 each time it is evaluated. Note that in such cases, Java will always round decimals *down*, so even a value of 4.9999 will be stored in integer format as 4.

RULE: if you wish to evaluate a real number expression, make sure that its arguments are also real numbers.

This is especially important for division. For example,

```
float answer;

answer = (1.0/2.0)+(1.0/2.0) ;[Conversion from double to float error]
answer = (1.0f/2.0f) + (1.0f/2.0f);      [ = 1.0 ]
answer = (1/2) + (1.0f/2.0f);            [ = 0.5 ]
answer = (1/2) + (1/2.0f);               [ = 0.5 ]
```

The same is true of variables that are evaluated, so make sure you remember what type of variables you are using in any arithmetic operation. Failure to do this is one of the more common (and frustrating) causes of errors in Java programming.

4.4 CONTROLLING MOVEMENT WITH LOOPS

So far, we have been able to predict how our programs will behave by finding the class with the `main()` method and reading down, line by line until we get to the end of the method. This may involve 'jumping' to methods in other classes, but within these methods, we also start at the top, reading down line by line until we get to the bottom.

This is not always the best way of doing things. Imagine creating a method that calculated the mean and variance of 200 numbers that had been typed in from the keyboard. Explicitly asking for, and adding up, 200 numbers, would be extremely tedious and inefficient to program.

In Java, (and most other procedural programming languages), it is possible to repeat parts of the code more than once. A repeated part of code is known as a *loop*. There are several different types of loop in Java. The rest of this section will investigate the more important ones.

4.4.1 The `for` Loop

The `for` loop has the following structure:

```
for (initialise; conditional expression; action)
{
    // Some code to be repeated
}
```

The section between the braces (which should be indented) is repeated several times depending on the three elements of the `for` statement.

For example,

```
int count;         // Loop counter.

for(count=0; count<16; count++)
{
    System.out.print(count+" ");
}
```

will display 0 1 2 3 4 5 6 7 8 9 10 11 12 13 14 15 on the screen.

Note that the `for` loop does not have a ';' at the end. Only commands within the loop are terminated with a semicolon. The loop itself should be enclosed by braces. The only exception is where the loop contains only one line, when the braces may be optionally omitted. The variable(s) used by the loop must be declared like any other, and can be modified like any other (even inside the loop). It is quite common to see the loop counter variable declared within the loop definition itself, often as a way of saving space. For example,

```
for(int count=0; count<16; count++)
    System.out.print(count+" ");
```

Have a look at Mean.java to see how the `for` loop may be used. Note the use of a loop 'counter', a common construct and very useful.

```
// ****************************************************************
/** Class that asks for a series of numbers and calculates their
  * mean. Demonstrates use of a loop to carry out a repeated task.
  * @author    Jo Wood.
  * @version   1.1, 30th May, 2001
  */
// ****************************************************************
```

```
public class Mean
{
    // ------------------ Object variables --------------------

    private int numSamples;      // Number of samples to be computed.

    private float total=0.0f,    // Total of all sample values.
                  mean =0.0f;    // Mean of the sample values.

    // --------------------- Methods ------------------------

    /** Asks user for number of samples and gathers input from
      * the keyboard. Note that input uses the KeyboardInput class
      * (written separately).
      */
    public void gatherInput()
    {
        // Create a new KeyboardInput object for input.
        KeyboardInput keyIn = new KeyboardInput();

        int count;          // Counts through samples.

        keyIn.prompt("How many samples? :");
        numSamples = keyIn.getInt();

        for (count=1; count <= numSamples; count++)
        {
            keyIn.prompt("Sample number " + count + " :");
            total += keyIn.getFloat();
        }
                    // Do the mean calculation.
        mean = total / numSamples;
    }


    /** Displays the results of sample mean calculations.
      */
    public void displayResults()
    {
        System.out.println("The mean of the " + numSamples +
                                        " samples is " + mean);
    }
}
```

This class contains several types of variable. The numSamples, mean and total are used by both methods (gatherInput() and displayResults()), so they are created as object variables outside of any method definition.

To make keyboard input as simple as possible, we have used a new KeyboardInput class. This is a class, specifically created for the examples in this book, that allows keys typed in on the keyboard to be stored in a variable. In keeping with the object-oriented philosophy, we are not particularly concerned with how this class works, just that it has some useful methods for us (prompt(), getInt() etc.).

Controlling Program Movement   81

We have used the `for` loop to ask repeatedly for keyboard input. The number of repetitions is controlled by the `numSamples` variable. The loop counts from 1 up to and including the number of samples.

Remember the code shown in `Mean.java` simply defines a class. If we wish to create an object from this class, we will need to either add a `main()` method, or (as shown below), create a separate class with a `main()` method that creates an object from `Mean`. Once an object has been created, we can invoke the program by calling the two methods:

```java
// ****************************************************************
/** Program to create a Mean object and call its input and output
 * methods.
 * @author    Jo Wood.
 * @version   1.1, 30th May, 2001
 */
// ****************************************************************

public class RunMean
{
    /** All java applications have a main() method.
     * This one creates a Mean object and calls its methods.
     */
    public static void main(String args[])
    {
        Mean mean = new Mean();
        mean.gatherInput();
        mean.displayResults();
    }
}
```

Running this class will create output similar to the following (but obviously dependent on what the user types in when prompted):

```
How many samples? : 4
Sample number 1 : 2.3
Sample number 2 : 3.3
Sample number 3 : 6.1
Sample number 4 : 1.0
The mean of the 4 samples is 3.175
```

4.4.2 The `do-while` Loop

The `do-while` loop has the following structure:

```
do
{
    // some code to be repeated
}
while (conditional expression);
```

The do-while loop repeats the section inside the braces while the condition at the bottom is satisfied.

For example,

```
float acidity;
KeyboardInput keyIn = new KeyboardInput();

System.out.println("Type in pH value (-999 to end)");

do
{
    keyIn.prompt(">");
    acidity = keyIn.getFloat();
}
while (acidity != -999);
```

The do-while loop should be used when (a) a loop counter is not needed (it was in Mean.java above); and (b) the code within the loop must be executed at least once regardless of the condition.

4.4.3 The 'while-do' Loop

This is very similar to the do-while loop except that the conditional statement is tested *before* the loop is entered rather than at the end. There is not actually a 'do' at the bottom, but it helps to think of it as an inverted do-while loop.

The structure of the 'while-do' loop is as follows:

```
while (conditional expression)
{
    // Some code to be repeated
}
```

The 'while-do' loop repeats the section inside the braces if the condition at the top is satisfied.

For example,

```
while ((percentage < 0) || (percentage >100))
{
    keyIn.prompt("% out of range, please correct:");
    percentage = keyIn.getInt();
}
```

The 'while-do' loop should be used when (a) a loop counter is not needed; and (b) the code within the loop may possibly never be executed depending on the condition being tested.

Controlling Program Movement

4.4.4 Nesting Loops

Loops of all types can themselves be placed inside other loops. This allows us to construct more sophisticated repetitions of tasks and is particularly useful for the manipulation of 2D and 3D spatial data structures. As an example, consider the following class to display a 2D grid of dots:

```
// ****************************************************************
/** Program that displays a regular 2D grid (up to 8x8 cells). It
 *  is intended to demonstrate how nested loops can be used to
 *  execute repeated tasks.
 *  @version  1.1, 31st May, 2001
 *  @author Jo Wood.
 */
// ****************************************************************

public class Grid
{
    // ------------------ Object variables --------------------

    private int gridSize= 0,      // Size of one grid dimension.
                maxSize = 8;      // Maximum grid size.


    // ---------------------- Methods ------------------------

    /** Asks for size of grid (cannot be greater than maxSize).
      */

    public void askGridSize()
    {
        // Create a new KeyboardInput object to handle input.
        KeyboardInput keyIn = new KeyboardInput();

        do
        {
            keyIn.prompt("Type in grid size (max. "+maxSize+") :");
            gridSize = keyIn.getInt();
        }
        while ((gridSize <= 0) || (gridSize > maxSize));
    }


    /* Display the grid on the screen.
     */

    public void displayGrid()
    {
        int row, col;                   // Grid coordinates.

        System.out.print("\n\t  ");     // Display top row.
        for (col=0; col<gridSize; col++)
            System.out.print(col + " ");


                                        // Display grid row by row.
```

```
            for (row=0; row<gridSize; row++)
            {
                System.out.print("\n\t" + row + " ");
                for (col=0; col<gridSize; col++)
                    System.out.print(". ");
            }
        }
    }
```

If a `Grid` object is created and its methods called, typical output might look like this:

```
Type in grid size (maximum of 8) : 25
Type in grid size (maximum of 8) : -4
Type in grid size (maximum of 8) : 7

          0 1 2 3 4 5 6
        0 . . . . . . .
        1 . . . . . . .
        2 . . . . . . .
        3 . . . . . . .
        4 . . . . . . .
        5 . . . . . . .
        6 . . . . . . .
```

In order to see how this technique can be used to help us model spatial information, we need to explore one further concept, that of the *array*.

4.5 GROUPING DATA IN ARRAYS

We have already considered one way in which Java can group items of data together by creating objects that contain a number of fields representing their state. This process is fundamental to object-oriented programming and we do this when our object-oriented model suggests some 'natural' grouping, such as a tree's height, age, and location. Typically, such grouping involves a relatively small number of component parts, often of different types. When gathered together in this way, an object will also typically show some kind of behaviour allowing the properties of the data to be changed in some way.

An alternative way of grouping information can be used when we wish to associate a larger number of items together all of the same type. An example of such a group is the *array*. An array is simply a group of variables all of the same type that can be treated as a single unit of a fixed size, but which allows individual members of the group to be manipulated. Loops provide one of the easiest ways of sequentially manipulating the contents of arrays.

4.5.1 Declaring and Initialising Arrays

To declare an array we do the following:

`type arrayName[] = new type[size];`

For example,

`int readings[] = new int[20];`

The example above will create an array of 20 integer variables collectively called `readings`. As with all declarations, all we have to do at this stage is to give the array a name and reserve enough memory to store potential values. Note that this declaration is similar in form to the creation of a new object from a class. The only difference being in the use of [square brackets].

Just as with any other type of variable, we can also separate our array declaration from its initialisation. For example,

```
int readings[];
readings = new int[20];
```

If all 20 integers have the same name (`readings`), how can we initialise each one individually? The answer lies in the fact that we can address each *element* in the array by an *index* inside square brackets. So, for example, if we wish to set the 5th item in the array to 24, we can do the following:

`readings[4] = 24;`

Why use a [4] as the index if it is the 5th element in the array we are interested in? We always index arrays by counting from 0 rather than 1. The reason for this is partly historical, deriving from languages where an array index was equivalent to the offset from the beginning of a list of numbers stored in computer memory.

A common way to initialise the contents of an entire array, will be to use a `for` loop with an index counter. For example,

```
for (int index=0; index<20; index++)
    readings[index] = 1;
```

Notice how we use the variable `index` (which goes from 0 to 19) to store the current array index as we pass through the list.

Occasionally, rather than initialising every element of an array with an identical number, we may wish to give each element separate values. This can be done at the same time as array declaration. For example,

`int ages[] = {23, 21, 22, 22, 29, 24, 30, 32};`

Note that the example above implicitly sets the array size to 8.

Arrays can be of any type of object as well as primitives. So, for example, we can create a forest model very easily using our `Tree` class as follows:

```
Tree forest[] = new Tree[400];    // Declare Tree array.

for (int tree=0; tree<400; tree++)
    forest[tree] = new Tree();
```

Note that the form of array declaration and initialisation is similar to that used for primitives, substituting the primitive type with the class name.

4.5.2 Multidimensional Arrays

The use of arrays described so far can be regarded as 'lists' of numbers or objects. The array index provides the position along the length of the list.

```
array[0]
array[1]
array[2]
array[3]
   :
```
etc.

However, it is also possible to create a 2D 'list' containing two index values:

```
array[0][0]  array[0][1]  array[0][2]  array[0][3]  ... etc.
array[1][0]  array[1][1]  array[1][2]  array[1][3]
array[2][0]  array[2][1]  array[2][2]  array[2][3]
array[3][0]  array[3][1]  array[3][2]  array[3][3]
   :
```
etc.

2D arrays are declared, initialised, and used in much the same way as their one-dimensional equivalent, the only difference being that two indices are required. For example,

```
int matrix[][] = new int[3][3];
```

will create a 3 by 3 array of integers called `matrix`. To set the middle value to 13 we can use the line

```
matrix[1][1] = 13;
```

The Raster Model

2D arrays are useful to us as they allow us to represent 2D spatial data as *rasters* (*bitmaps* in computer graphics parlance). Images such as photographs and those produced from satellite remote sensing are conveniently modelled as rasters where each value in the 2D array stores a 'colour' value (see Figure 4.1). Rasters used for spatial modelling also need to be *georeferenced*. A raster model should store sufficient information to identify the location and extent of an area on the ground. One way of doing this is to store the dimensions represented by a raster cell (known as its *resolution*) along with the location of the raster's corner or *origin* (see Figure 4.2).

Figure 4.1 A raster image showing the arrangement of raster cells.

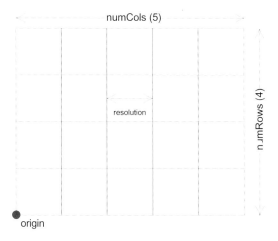

Figure 4.2 A simple raster model of space.

We can use a 2D array and nested loop to model georeferenced raster maps very easily. `RasterMap` shown below illustrates how we might store the necessary information in a single class.

```
import java.awt.geom.*;         // For Point2D class.
//    ****************************************************************
/** Models a GIS raster map. Stores the raster resolution and
  * dimensions as well as the raster values.
  * @author Jo Wood
  * @version 1.2, 8th September, 2001
  */
//    ****************************************************************

public class RasterMap
{
    // ---------------- Object variables -----------------

    private float[][] raster;              // Raster values.
    private float     resolution;          // Raster resolution.
    private int       numRows,numCols;     // Dimensions of raster.
    private Point2D   origin;              // Origin of raster.


    // ----------------- Constructor --------------------

    /** Creates a raster map with the given dimensions.
      * @param numRows Number of rows in raster.
      * @param numCols Number of columns in raster.
      * @param resolution Size of 1 side of a raster cell.
      * @param origin Locatio of bottom left corner of raster.
      */
    public RasterMap(int numRows, int numCols,
                     float resolution, Point2D origin)
    {
        // Store incoming messages as object variables.
        this.numRows     = numRows;
        this.numCols     = numCols;
        this.resolution  = resolution;
        this.origin      = origin;

        // Declare and initialise new raster array.
        raster = new float[numRows][numCols];

        for (int row=0; row<numRows; row++)
            for (int col=0; col<numCols; col++)
                raster[row][col] = 0.0f;
    }
}
```

The constructor uses a new Java keyword `this`. It allows us to distinguish between the incoming message parameters (`numRows` and `numCols`) and the object variables (`this.numRows`, `this.numCols`). We shall see several more uses of `this` in later chapters, but for the moment you can regard the word as meaning 'the object in which this line of code appears'.

The values of the raster are stored in the 2D floating point array `raster[][]`. In the 2D case, array rows are indexed before array columns. This is the opposite way round to normal Cartesian coordinate referencing, but similar to matrix (and remote sensing) conventions.

`RasterMap` uses a new class in Java called `Point2D` to store the geometric origin of the raster. In order to use this, we need to include the line `import awt.geom.*;` which makes the class available to our program. This class stores a pair of values representing the georeferenced bottom-left corner of the raster. We shall explore some of the other classes that form part of the `java.awt.geom` package in Chapter 7 when we consider an alternative form of spatial representation – the vector model.

Combining the origin information and the raster resolution allows us to *transform* between the coordinate system used to reference the array (integer indices ranging from (0,0) to (numRows-1, numCols-1) and the georeferencing system used to locate objects 'on the ground' (floating point values in (easting, northing) order). So, for example, we might add a method to our class that reports the attribute of the raster at the given georeferenced coordinates:

```
/** Reports the attribute of the raster at the given georeferenced
  * coordinates.
  * @param easting Easting of the location to query.
  * @param northing Northing of the location to query.
  * @return attribute of the raster at query location.
  */
public float getAttribute(float easting, float northing)
{
    // Transform georeferenced coords into raster coords.

    int col = (int)((easting - origin.getX())/resolution);
    int row = (numRows-1) -
              (int)((northing - origin.getY())/resolution);

    return raster[row][col];
}
```

At the heart of this method, the transformation from georeferenced coordinates (`easting, northing`) to raster indices (`row, col`) involves translating the raster origin to (0, 0) and calculating how many units of resolution are required to move to the query location. Note that this calculation could produce an array index that was 'out of bounds' if the query location falls outside the area covered by the raster.

4.6 LAYING OUT GRAPHICAL COMPONENTS

In the previous chapter we saw how to add a simple label to a window using the `add()` method of a `contentPane`. In that example there was little to do other than place a single component in a window and resize it appropriately. Once we start to create more sophisticated user interfaces with many components, we need to be able to exercise a little more control over their arrangement within a window.

We can ask the AWT to handle laying out components in such a way that all graphical objects are positioned automatically and 'shuffled about' if the window is resized. As an example, consider the following class for displaying three components:

```java
import javax.swing.*;          // For Swing components
import java.awt.*;             // For AWT components.

// ***************************************************************
/** Creates a simple Swing-based window and demonstrates how Java
  * lays out graphical components.
  * @author    Jo Wood.
  * @version   1.1, 17th June, 2001
  */
// ***************************************************************

public class SimpleWindow2 extends JFrame
{
    // ----------------- Object Variables --------------------

    private Container contentPane;    // This is the container into
                                      // which components are added.


    // ------------------- Constructor -----------------------

    /** Creates a simple window with an equally simple label
      * inside.
      */
    public SimpleWindow2()
    {
        // Create the window with a title.
        super("Another simple window");
        setDefaultCloseOperation(WindowConstants.DISPOSE_ON_CLOSE);
        contentPane = getContentPane();

        // Set the style of layout.
        contentPane.setLayout(new FlowLayout());

        // Create graphical components.
        JLabel label1 = new JLabel("A simple label");
        JLabel label2 = new JLabel("Another simple label");
        JButton button = new JButton("Press Me");

        // Add graphical components to the window.
        contentPane.add(label1);
        contentPane.add(button);
        contentPane.add(label2);

        // Get Java to size the window and make it visible.
        pack();
        setVisible(true);
    }
}
```

When an object is created out of SimpleWindow2, a window like the one shown in Figure 4.3(a) is created. Without explicitly using any coordinates, the program calculates how big the window should be and resizes accordingly. More importantly, if the user resizes the window, the components within it are readjusted to fit within it (Figure 4.3(b)).

Figure 4.3 A simple window with three components controlled by the layout manager.

SimpleWindow2 is similar to our first window example considered in Section 3.4.2 but with the following important differences. First, we have created three components to add to the window in order to demonstrate how these are rearranged by Java. We have created two labels and one button. Note that the order that the label and button objects are created does not necessarily determine their position within the window. It is the order in which they are *added* using the add() method.

Second, we have instructed Java to use a *layout manager* to take control of positioning the components. The layout manager is determined by calling the setLayout() method of the content pane. The Java AWT has several useful layout managers that can be invoked in your container classes. These are summarised below.

4.6.1 Flow Layout

This is one of the simplest layout managers to use. Each time a new component is added, the manager positions it to the right of the last one that was added. If the width of the container prevents fitting the new component to the right, it will start adding new components below existing ones (Figure 4.4).

Figure 4.4 Five buttons arranged using a FlowLayout manager.

4.6.2 Border Layout

This is the default layout scheme for the `JFrame`'s content pane and allows up to five components to be added (4 around the edge and 1 in the centre, see Figure 4.5). When adding a component use

`add(component,position);`

where `position` is one of
      `BorderLayout.NORTH,`   `BorderLayout.SOUTH,`
      `BorderLayout.WEST,`   `BorderLayout.EAST` or
      `BorderLayout.CENTER` (note the 'ER' spelling).

This is the only layout manager where the order of `add()` calls, makes no difference to the positioning of components since they are overridden by the `position` message.

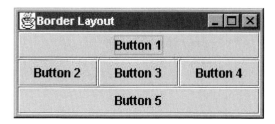

Figure 4.5 Five buttons arranged using a `BorderLayout` manager.

4.6.3 Grid Layout

This is a useful layout if components of similar size need to be displayed at regular intervals. When creating the `GridLayout` object, use the constructor that allows the number of rows and columns to be specified.

Figure 4.6 Five buttons arranged using a `GridLayout` manager.

The example shown in Figure 4.6 was created with the line

```
new GridLayout(5,1)
```

Components are added in column order starting at the top row working down. Note that all components are resized to be the same as the single largest component.

4.6.4 Other Layout Managers

Java contains other layout managers that are a little more complicated to use, but provide further flexibility in the way they organise graphical components on the screen.

The `GridBagLayout` manager allows different components to be positioned in different ways with some taking priority over others. Unfortunately, programming with the `GridBagLayout` is rather complicated.

The `CardLayout` manager allows several 'virtual layouts' to be layered on top of each other. By selectively making each layout visible, it is possible to create a 'tabbed panel' type effect. Again using this manager can be complicated, and if you wish to create a tabbed pane, it is suggested you use the `JTabbedPane` component instead.

4.7 CASE STUDY: CREATING A DISPLAYABLE RASTER MAP

The `RasterMap` class created in Section 4.5.2 can store spatial data in raster form, but as yet it cannot be displayed on screen. This case study shows how a simple graphical user interface can be created to display a `RasterMap`. It demonstrates some of the principles of graphical user interface design as well as introducing some new Java functionality for displaying images.

4.7.1 Designing the Graphical User Interface

A graphical user interface (GUI, pronounced 'gooey') provides a connection between the internal workings of a program and our interaction with it. Good GUI design is a complex and subtle skill which we can only touch on here, but in considering the design of a GUI for our `RasterMap`, some of the more important and practical aspects of GUI design will be made clear.

The first and probably most important point is that graphical representation within a program should be kept as separate as possible from its storage and analysis elements. This allows both to be developed independently and aids program clarity. In Java programming terms, this principle usually means that you will develop separate methods or entire classes for GUIs.

The program objective of this case study is to create a suitable GUI that will display the contents of a raster map graphically along with its supplementary information such as title, resolution and dimensions. We already have a class `RasterMap` that represents this information, so our task becomes one of creating a new class to handle the GUI issues.

Before starting to program a GUI, it is worth sketching out some possible graphical designs. It is much simpler to change these designs than it would be to alter a completed Java program and it also gives us some idea of how we should approach the construction of the GUI. Figure 4.7 shows one such design, which we will use as the basis of this case study. The design is simple and hopefully familiar to users of windowing systems (for example, standard window manipulation controls in the top-right corner, centred 'OK' button at the bottom of the window).

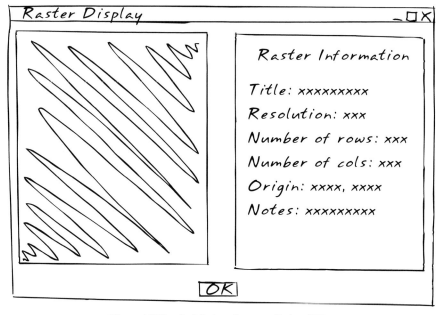

Figure 4.7 Sketched design of a raster display GUI.

You may also notice that its structure is somewhat hierarchical in that the window 'container' has within it a graphical panel container, a text container and a button. The text panel on the right itself contains seven further text objects. This gives us a clue as to how we might approach programming the GUI. Each graphical 'object' can be represented by a Java object, which itself may be composed of further Java objects. We can then use Java's layout managers to arrange the objects within a window for us.

4.7.2 Creating a RasterMap Panel

The text and button elements of the GUI design are relatively straightforward as they can use classes such as `JLabel` and `JButton` that we have already come across. The display of the raster on the left of the window presents us with a more complex programming problem as there is no existing Java class that will do this for us in one simple step. To keep things simple and object-oriented, we will create a new class specifically for displaying rasters graphically. The basis for this class will be the Swing class `JPanel`. This is an invisible container, into which we can draw graphics or place other graphical components. As we shall see in the next section, `JPanel`s can be very useful when arranging components within a window. The outline structure of our new class is shown below.

```
import javax.swing.*;        // For Swing components.
import java.awt.*;           // For graphics context.
//    ***************************************************************
/** Panel that draws a GIS raster map.
  * @author Jo Wood.
  * @version 1.0, 8th September, 2001
  */
//    ***************************************************************
public class RasterPanel extends JPanel
{
    // --------------------- Constructor -----------------------

    /** Creates a panel and draws the given raster upon it.
      * @param rasterMap Raster map to draw.
      */
    public RasterPanel(RasterMap rasterMap)
    {
        super();                       // Call JPanel's constructor.
        setRasterMap(rasterMap);       // Store raster map.
    }

    // --------------------- Methods ---------------------------

    /** Sets a new raster map to draw.
      * @param rasterMap New raster map to draw.
      */
    public void setRasterMap(RasterMap rasterMap)
    {
        // Code for storing graphical version of raster map here.
    }

    /** Draws graphics on the panel.
      * @param g Graphics context in which to draw.
      */
    public void paintComponent(Graphics g)
    {
        // Code for drawing raster map here.
    }
}
```

We extend `JPanel` in order to give our class all the behaviour of a `JPanel` container. The constructor of `RasterPanel` expects a `RasterMap` object as an incoming message which it will then store using the method `setRasterMap()`. This public method will also allow us to change the raster map displayed from within other classes. The method `paintComponent()` is one that we have not yet seen but is defined for all `JComponents` (of which `JPanel` is one). This is the method that draws the component on screen. The method takes a single incoming parameter, a *graphics context*, which is by convention given the variable name g and is an object of type `Graphics`. A graphics context tells Java where to draw graphics as well as containing many useful methods for drawing graphical features. In the case of a `JPanel`, this simply consists of drawing a blank area in the default window colour. Our class will *override* the `paintComponent()` method so that we can force it to draw the raster on screen instead.

Images and Colour Transformations

In order to display a raster graphically, we need to perform some kind of *transformation* that converts the numerical value in each raster cell into some colour that represents it. To keep things simple, we will perform a very simple transformation that converts all raster values between 0 and 255 into a grey scale value ranging from black (0) to white (255). Values outside of this range will have undefined colours. The result will be a grey-scale raster image much as was shown in Figure 4.1.

We can store the raster image in a Java class, `Image`, which forms part of the AWT. The method `setRasterMap()` shows how this can be achieved.

```
// ------------------- Object variables -------------------

private Image rasterImage;        // Raster map to display.


// ---------------------- Methods ------------------------

/** Sets a new raster map to draw. Currently assumes raster values
  * are scaled between 0 and 255.
  * @param rasterMap New raster map to draw.
  */
public void setRasterMap(RasterMap rasterMap)
{
    int numRows = rasterMap.getNumRows();
    int numCols = rasterMap.getNumCols();

    // Transfer raster values into pixel array.
    int pixels[] = new int[numRows * numCols];
    int pixelCount=0;

    for (int row=0; row<numRows; row++)
    {
        for (int col=0; col<numCols; col++)
        {
            int greyLevel = (int)rasterMap.getAttribute(row,col);
            Color colour = new Color(greyLevel,greyLevel,greyLevel);
```

```
                pixels[pixelCount++] = colour.getRGB();
        }
    }

    // Create a drawable image out of pixel array.
    rasterImage = createImage(new MemoryImageSource(
                                numCols,numRows,pixels,0,numCols));
    repaint();
}
```

To convert our `RasterMap` into an `Image`, we need to do two things. First, the `Image` class expects a one-dimensional array containing all pixel values. Since the `RasterMap` stores cell values as a 2D array, we use two loops, one nested inside the other, to convert from the rectangular raster to the linear pixel array. Second, we need to convert raster values into their grey-scale colour value. This can be achieved using the Java class `Color`, which contains a method `getRGB()` that will perform the conversion for us. This combined transformation is shown in Figure 4.8. Finally, the image is created using the method `createImage()` that is inherited from `JPanel`.

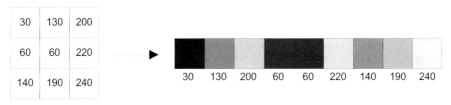

Figure 4.8 A 2D raster to 1D image transformation.

The final task in creating a `RasterPanel` is to draw the newly created image on screen. We do this by overriding the `paintComponent()` method of `JPanel` and replacing it with our own image drawing code.

```
/** Draws graphics on the panel.
  * @param g Graphics context in which to draw.
  */
public void paintComponent(Graphics g)
{
    super.paintComponent(g);    // Paint background.

    // Scale and draw image to fit inside panel.

    Dimension panelSize = getSize();

    g.drawImage(rasterImage,0,0,
                panelSize.width,panelSize.height,this);
}
```

The paintComponent() method contains an incoming message which is an object created from the Graphics class inside the AWT. By convention, the object name is given the letter g, even though a more explanatory name such as graphics might be more useful. The Graphics class contains a number of useful methods for drawing on the screen, but we only need one of them – drawImage() to draw our raster image. This method takes six parameters that control where and how the image is drawn. The first is the image to display and the following two are the x and y coordinates of the top left corner of the image position in the panel. The third and fourth parameters give the dimensions of the image, which, in our example, we set to the size of the panel. This means that if the panel is resized by the user (or window manager), the image will be resized accordingly. The final parameter is the object in which the image display will take place.

4.7.3 Arranging Graphical Components

Once we have created a RasterPanel class for displaying rasters, we can place it within a window along with the supporting text describing the raster map. The overall structure of the class to do can be similar to those we have already seen in Section 4.6, the only difference being that the arrangement of components is a little more complicated.

To arrange the components as shown in Figure 4.7, we can create a number of containers, each of which contain either further containers or graphical components. Figure 4.9 shows a suitable hierarchical arrangement of panels and components.

Inside the BorderLayout of JPanel's contentPane is placed the OK button to the south and the remaining containers in the centre. The main display of the window contains two equally sized areas, one containing our newly created RasterPanel, the other a grid of JLabels describing the raster.

The use of multiple and nested layout managers is quite common in Java GUI design, a process that is made simpler by sketching the GUI arrangement before commencing programming. Even with the relatively simple grid, flow and border layouts, it is possible to create quite sophisticated arrangements of components by nesting different layout managers within each other.

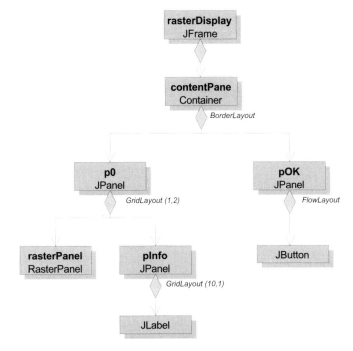

Figure 4.9 Hierarchical arrangement of graphical components in the RasterDisplay GUI. Objects are labelled in bold, classes in a plain font, and layout managers in italics.

The code for the RasterDisplay class is shown below.

```
import javax.swing.*;         // For Swing components
import java.awt.*;            // For AWT components.
import java.awt.geom.*;       // For Point2D class.

// ****************************************************************
/** Creates a simple window for displaying raster maps.
  * @author    Jo Wood.
  * @version   1.2, 9th September, 2001
  */
// ****************************************************************

public class RasterDisplay extends JFrame
{
    // ---------------------- Main method ----------------------

    /** Creates an application that launches the raster display.
      * @param args Command line arguments (ignored).
      */
    public static void main(String[] args)
    {
        new RasterDisplay();
    }
```

```java
    // ---------------------- Constructor ----------------------

    /** Creates a top-level window and simple raster and displays
     * it on its own panel.
     */
    public RasterDisplay()
    {
        // Create the window with a title.
        super("A simple raster display");
        setDefaultCloseOperation(WindowConstants.DISPOSE_ON_CLOSE);
        Container contentPane = getContentPane();

        // Place everything inside a panel with spacing border.
        JPanel p0 = new JPanel();
        p0.setLayout(new GridLayout(1,2));
        p0.setBorder(BorderFactory.createEmptyBorder(4,4,4,4));

        // Create a simple raster map to display in left of panel.
        RasterMap rastMap = new RasterMap(30,30,50,
                                    new Point2D.Float(33,75));

        RasterPanel rastPanel = new RasterPanel(rastMap);
        rastPanel.setPreferredSize(new Dimension(200,200));
        rastPanel.setBorder(BorderFactory.createEtchedBorder());
        p0.add(rastPanel);


        // Display raster information on right of the panel.
        JPanel pInfo = new JPanel(new GridLayout(10,1));

        pInfo.setBorder(BorderFactory.createEtchedBorder());
        pInfo.add(new JLabel("Raster Information", JLabel.CENTER));
        pInfo.add(new JLabel());
        pInfo.add(new JLabel("  Title: (not yet defined)"));
        pInfo.add(new JLabel("  Resolution: " +
                                rastMap.getResolution()));
        pInfo.add(new JLabel("  Number of rows: " +
                                rastMap.getNumRows()));
        pInfo.add(new JLabel("  Number of cols: " +
                                rastMap.getNumCols()));
        pInfo.add(new JLabel("  Origin: " +
                                rastMap.getOrigin().getX()+","+
                                rastMap.getOrigin().getY()));
        pInfo.add(new JLabel("  Notes: (not yet defined)"));
        p0.add(pInfo);

        // Add the raster panel and an ok button.
        contentPane.add(p0, BorderLayout.CENTER);

        JPanel pOK = new JPanel();
        pOK.add(new JButton("OK"));
        contentPane.add(pOK,BorderLayout.SOUTH);

        // Get Java to size the window and make it visible.
        pack();
        setVisible(true);
    }
}
```

Some fine tuning of the arrangement of components is achieved with the use of a BorderFactory to create a four pixel border around the two main areas of the window. The OK button is also placed inside its own panel (pOK). Without this, the layout manager will fill the entire 'south' of the window with the button. A sample of output from the program is illustrated in Figure 4.10 showing the display of a small raster with random values between 0 and 255.

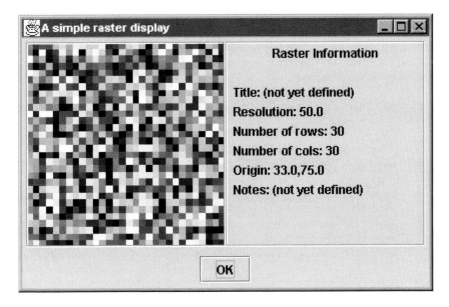

Figure 4.10 The final raster display window.

4.8 SUMMARY

This chapter first considered how we can manipulate the contents of variables by the use of operators. These include arithmetic operators (+ - / * %) and logical operators (&& || == != < <= >=). To avoid confusion, complex expressions that involve a number of arithmetic or logical operators should use brackets and spaces to make the order in which the expression is calculated clear. To avoid unintentional rounding of numbers, care should be taken to make sure all floating point calculations contain floating point arguments.

The evaluation of arithmetic and logical operations can be used to force parts of program code to be repeated in a loop. The use of looping structures can significantly increase the efficiency with which we get the computer to perform repeated tasks. They become particularly powerful when loops are nested within each other.

Collections of similar primitive or object types can be gathered together in arrays, which, when combined with looping structures, allow large volumes of data to be processed efficiently. We considered how 2D arrays can be used to hold spatial data in the form of the raster model.

Finally, the case study demonstrated how we can add a graphical 'front-end' to a program in order to communicate effectively with the user. The design of graphical user interfaces involved first sketching out a likely design, then translating that design into a hierarchy of graphical components, before coding them using Java's Swing and AWT graphical toolkits.

Three types of loops have been discussed so far, the `for` loop, the `do-while` loop and the 'while-do' loop. Loops can be made to perform quite powerful tasks by nesting one inside another.

By the end of this chapter and associated exercises on the web, you should be able to

- create an arithmetical expression involving numerical variables;
- understand how values may be rounded and the order in which parts of an arithmetic or logical expression is evaluated;
- know which of the three types of loop in Java should be used for a given programming task involving repetition;
- predict the number of times a given line of code inside a loop or nested loops would be executed;
- understand how spatial data may be modelled using a raster data structure; and
- design and implement a simple graphical user interface using the Swing toolkit.

CHAPTER FIVE

Making Decisions

We develop the idea of controlling program flow in this chapter by considering how we can execute sections of code depending on conditional statements. This allows us to code decision-making processes in our Java programs. The ability to make decisions in Java (and any programming language) vastly increases the effectiveness of the language in being able to model real-world behaviour.

The process of encoding decision-making is illustrated with the development of a set of classes for representing spatial objects and identifying the relationships between them.

5.1 MAKING DECISIONS WITH IF

5.1.1 The Simple if Statement

If we are modelling some kind of behaviour, the actions performed as part of that behaviour frequently depend on some set of conditions.

For example,

To boil some water in a kettle:

Check water level.
If water level is too low, fill up with water.
Turn kettle on.

Likewise, we can make tasks in Java methods dependent on some condition by using the if statement.

The if construction has the following structure:

```
if (conditional expression)
{
    // Some code that is dependent on the condition
}
```

The conditional expression has exactly the same form as that used in the for and while loops discussed in the previous chapter. The same rules for using semicolons, braces and indentation also apply. That is, no semicolon is placed at the end of the `if` line; braces are required if more than one conditional action is performed; and all conditional actions should be indented.

For example,

```
if (number1 > number2)
{
    System.out.println("Swapping numbers.");

    int temp = number1;
    number1  = number2;
    number2  = temp;
}
```

In this example we have declared and initialised a variable `temp` inside `if`'s braces. This variable is known as a *local variable* since it is only visible to lines of code inside the pair of braces that contain it. Creating local variables is quite common when the variable is only needed if a conditional expression is true.

The code above compares two variables and swaps their contents if the first number is larger than the second. The result of this decision-making process is to guarantee that the value of `number1` will never be larger than `number2`. We can use this process as the basis of a more sophisticated list sorting process that places the contents of an array into numerical order:

```java
// ******************************************************************
/** Class to sort a list of numbers into numerical order. Uses a
  * 'bubble sort' to demonstrate the use of decision making.
  * @author   Jo Wood.
  * @version  Version 1.1, 10th June, 2001
  */
// ******************************************************************

public class Sort
{
    // ------------------ Constructor --------------------

    /** Creates a list of numbers and sorts them.
      */
    public Sort()
    {
        int list[] = {17,3,60,128,11,11,2};
        sortList(list);
    }

    // -------------------- Methods ----------------------

    /** Sorts the given list into numerical order. Displays the
      * numbers in the list before, during and after the ordering.
      * @param list List of numbers to sort.
      */
    public void sortList(int[] list)
    {
```

```
            boolean listChanged;    // Records whether list has changed.
            int temp;               // Temporary storage.

            do
            {
                displayList(list); // Display current state of list.
                listChanged = false;

                for (int i=0; i<list.length-1; i++)
                {
                    if (list[i] > list[i+1])
                    {
                        temp     = list[i];
                        list[i]  = list[i+1];
                        list[i+1] = temp;
                        listChanged = true;
                    }
                }
            }
            while (listChanged == true);
        }

        // ---------------- Private Methods ------------------

        /** Displays the given list of numbers.
          * @param list List of numbers to display.
          */
        private void displayList(int list[])
        {
            for (int i=0; i<list.length; i++)
                System.out.print(list[i] + " ");

            System.out.println();
        }
    }
}
```

The heart of the program is the method `sortList()` that takes an array of integers as an incoming message and attempts to sort them in numerically increasing order. It is based on the `if` comparison that considers swapping a pair of variables if the first is greater than the second. We use a simple `for` loop to compare all values in the list in this way.

```
for (int i=0; i<list.length-1; i++)
{
    if (list[i] > list[i+1])
    {
        temp     = list[i];
        list[i]  = list[i+1];
        list[i+1] = temp;
    }
}
```

Yet passing through a list just once like this does not guarantee that the array will be sorted. So, how many times do we need to repeat this process before we can be sure that all the numbers in the array are in numerical order? The answer depends on the state of the list before it is sorted. Clearly, if the numbers in the list were already in numerical order, we would not need to perform any variable swapping. If the numbers were in opposite order, we would have to perform many more swaps.

One way of achieving an efficient swapping process is to create a loop that keeps swapping pairs of numbers until there is no more swapping to do. We can store whether or not further swapping might be required in a `boolean` variable (`listChanged`) that is set to `true` whenever a pair of numbers are found that are out of sequence. The list will be sorted when, after comparing all numbers in the list, `listChanged` remains `false`.

The result of running this program will give the output as follows:

```
17  3 60 128 11 11   2
 3 17 60  11 11  2 128
 3 17 11  11  2 60 128
 3 11 11   2 17 60 128
 3 11  2  11 17 60 128
 3  2 11  11 17 60 128
 2  3 11  11 17 60 128
```

Sorting Algorithms

Sorting data into some known order is an important part of the processing and efficient handling of data. Many processes that act on groups of data perform much more effectively if those data are sorted in some way.

The type of sorting used in the example above is known as a *bubble sort* as numbers that are out of sequence appear to 'bubble' up to their correct position in the list. This is one of many hundreds of different sorting *algorithms* that have been developed by computer scientists in the last few decades. The most appropriate type of sorting algorithm depends on many factors such as the size of the list to sort, the nature of the unsorted data, and way in which data are added or removed from a sorted list.

For more details on sorting algorithms, see Gosling *et al.* (2001).

We will consider one further example of decision-making using the `if` statement, which we will develop in the next sections of this chapter. The goal of this example is to provide imaginary support for a teacher who wishes to convert a series of numerical percentage marks into their equivalent letter grades (such as C+, B– etc.). The code below shows how this might be achieved:

Making Decisions

```
//   ****************************************************************
/** Class to take the grind out of marking students' exams.
  * Converts between percentage and letter grades.
  * @author Jo Wood
  * @version  1.1, 3rd June, 2001
  */
//   ****************************************************************

public class Exam
{
    // -------------------- Constructor ---------------------

    /** Initialises the Exam class by asking for a percentage value
      * and reporting its letter equivalent.
      */
    public Exam()
    {
        int percGrade = askPercGrade();
        System.out.println("The grade for " + percGrade + " is " +
                           percToLetter(percGrade));
    }

    // ---------------------- Methods -----------------------

    /** Asks for percentage to convert into a letter grade.
      * @return Percentage typed in by user.
      */
    public int askPercGrade()
    {
        int percent;          // Percentage value to convert.

        // Create a new KeyboardInput object to handle input.
        KeyboardInput keyIn = new KeyboardInput();
        do
        {
            keyIn.prompt("Type in an exam percentage: ");
            percent = keyIn.getInt();
        }
        while ((percent < 0) || (percent > 100));

        return percent;
    }

    /** Converts an exam percentage into a letter grade (using
      * a rather mean marking scheme).
      * @param Percentage grade to convert.
      * @return Letter grade equivalent to the given percentage.
      */
    public String percToLetter(int percGrade)
    {
        // Initialise grade text.
        String letterGrade = new String();
        String modifier    = new String();

        // Find the correct letter.
        if (percGrade < 40)
            letterGrade = "F";
        if ((percGrade >= 40) && (percGrade < 50))
            letterGrade = "E";
```

```
            if ((percGrade >= 50) && (percGrade < 60))
                letterGrade = "D";
            if ((percGrade >= 60) && (percGrade < 70))
                letterGrade = "C";
            if ((percGrade >= 70) && (percGrade < 80))
                letterGrade = "B";
            if (percGrade >= 80)
                letterGrade = "A";

            // Add the + or - modifier.
            if (percGrade%10 < 3)
                modifier = "-";

            if (percGrade%10 > 6)
                modifier = "+";

            // All marks below 30 will be F-, all above 90 will be A+
            if (percGrade < 30)
                modifier = "-";
            if (percGrade >= 90)
                modifier = "+";


            // Add the modifier to the grade.
            letterGrade += modifier;

            return letterGrade;
    }
}
```

The class stores information about letters as well as numbers, so we use a new type of variable, the String designed for storing and manipulating text. The String class is built into the Java language and can be used at any point in your own classes. Note also how use is made of message passing with the methods percToLetter() and askPercGrade(). By placing these behaviours in their own methods, we can call them at any point from other classes. The % (modulus) operator is used to extract the last digit from the percentage value. This allows us to assign a + or – to the letter grade depending on whether the percentage is towards the upper or lower end of the grade.

The result of creating an Exam object several times might be as follows (of course depending on what the user types in)

```
Type in a percentage grade : 57
The grade for 57 is D+

Type in a percentage grade : 82
The grade for 82 is A-

Type in a percentage grade : 65
The grade for 65 is C
```

5.1.2 The `if-else` Construction

It should be apparent from the Exam class above that there is a degree of inefficiency in the decision-making. If, for example, a mark is deemed to be an F grade, there is no need to test to see whether it might also be an E, D, C, B or A grade. In other words, the decisions to be made are *mutually exclusive*. If one condition is satisfied, we know the others will not be. Java has the facility to model mutually exclusive decisions using the `if-else` construction.

```
if (conditional expression)
{
    // Some code dependent on the condition being true
}
else
{
    // Some code dependent on the condition being false
}
```

For example,

```
if (balance < 0)
    System.out.println("You are overdrawn.");
else
    System.out.println("Would you like to take out a loan?");
```

The same rules apply to braces, indentation and the use of semicolons as it does for normal `if` statements. It is important that indentation is used so that program flow can be followed easily. If the code is not laid out correctly, it is very easy to get lost. This is especially true for the 'one-line' versions of the `if` clause (as shown above) that does not use braces to identify the conditional action.

5.1.3 Nested `if` Statements

As we try to model more complex behaviour in our Java programs, it is possible to make decisions only if previous decisions have been made. In other words, we can nest `if` statements inside each other. When we do this, it is essential that we use the correct indentation so that we can follow program flow easily. For example,

```
if (dark == true)
    System.out.println("The park is closed.");
else
{
    System.out.print("The park is open ");
    if (weekend == true)
        System.out.println("for football and perambulation.");
    else
        System.out.println("for your enjoyment.");
}
```

Nesting `if` and `else` commands can increase program efficiency if thought about carefully. For example, consider how the `percToLetter()` method in our Exams class can be improved:

```java
/** Converts an exam percentage into a letter grade (using
 * a rather mean marking scheme).
 * @param Percentage grade to convert.
 * @return The letter grade equivalent to the given percentage.
 */
public String percToLetter(int percGrade)
{
    // Initialise grade text.
    String letterGrade = new String();

    // Find the correct letter.
    if (percGrade < 40)
        letterGrade = "F";
    else
        if (percGrade < 50)
            letterGrade = "E";
        else
            if (percGrade < 60)
                letterGrade = "D";
            else
                if (percGrade < 70)
                    letterGrade = "C";
                else
                    if (percGrade <80)
                        letterGrade = "B";
                    else
                        letterGrade = "A";

    // Add the + or - modifier.
    if ((percGrade%10 < 3) || (percGrade < 30))
        letterGrade += "-";
    else
        if ((percGrade%10 > 6) || (percGrade >=90))
            letterGrade += "+";
    return letterGrade;
}
```

Note that while the nested indentations might seem a little odd at first, the vertical alignment of `if` and `else` pairs helps us to breakdown the logic of the operation into smaller more understandable steps.

> *There is a subtle problem with the logic of this new method. In some circumstances, the method will fail to report the correct grade. Try to spot the mistake (and suggest a remedy for it), by* dry running *the program. That is, work through it line by line following what would happen if various percentage values were passed to the method. Which values produce incorrect results?*

5.1.4 When to Use `if` Statements

The `if` statement should be used for

- Simple conditions
- Nested conditions
- Conditions that test for a range of values.

The `if-else` constructions should be used for

- Mutually exclusive actions.

However, if you find yourself writing a method containing tens of consecutive `if` statements, it is quite likely that there is a more efficient and object-oriented solution to the behaviour that you are trying to represent.

5.2 MAKING DECISIONS WITH SWITCH

5.2.1 The `switch` Statement

Sometimes a task may require a host of actions to be performed depending on many specific conditions. An example might be a menu system where eight different actions might be possible depending on which of the keys A–H are pressed.

In such cases, we can use a slightly more compact alternative to the `if` statement, called the `switch` construction. It has the following general structure:

```
switch (expression)
{
    case (constant1):
        // Code dependent on expression being equal to constant1
        break;

    case (constant2):
        // Code dependent on expression being equal to constant2
        break;

    case (constant3):
        // etc.
}
```

The expression is evaluated in the line with the `switch` statement, and if it matches any of the constants identified by the `case` lines, the code associated with that constant is executed. For example,

```
switch(age)
{
    case (5):
        System.out.println("You can now go to big school.");
        break;

    case (14):
        System.out.println("You can now visit pubs by yourself.");
        break;

    case (16):
        System.out.println("You can now join the army.");
        break;

    case (18):
        System.out.println("You can now vote.");
        System.out.println("You can now drink in pubs.");
        break;
}
```

5.2.2 Flow-through, Breaks and Defaults

`case` statements within a `switch` construction can always be replicated by `if`s and `if-else`s. They are used because they can make the flow through a program more obvious to the reader of the program.

The `break` statement after each case is vital. Whenever a `break` is reached, program flow moves to the end of the `switch` construction (i.e. to the line after the `switch`'s closing brace). If a `break` line is not included for any particular case, flow drops through to the next `case` regardless of whether it matches the given constant. The use of `break` is similar, but not identical, to putting an `else` after an `if` statement. In other words, it makes each case *mutually exclusive*.

Sometimes you might also want a default action to be performed if none of the cases is met. This can be done by adding a `default:` at the bottom of the `switch`.

```
default:
    // Code dependent on no cases being met.
    break;
```

For example, to our example above we might add,

```
default:
    System.out.println("Not much happening this year.");
    break;
```

For a more useful example, we can add a couple of methods to our `Exam` class that asks for a letter grade and converts it back into a percentage (for simplicity we shall ignore the + or − modifier, and assign the median percentage mark to each class).

```java
/** Asks for a letter grade to convert into a percentage.
  * @return Letter grade typed in by user.
  */
public String askLetterGrade()
{
    // Create a new KeyboardInput object to handle input.
    KeyboardInput keyIn = new KeyboardInput();

    // Initialise grades
    String letterGrade;

    do
    {
        keyIn.prompt("Type in a letter grade :");
        letterGrade = keyIn.getString();
    }
    while ((letterGrade.compareTo("A") < 0) ||
           (letterGrade.compareTo("F") > 0) ||
           (letterGrade.length() > 1));

    return letterGrade;
}
```

```java
/** Converts a letter grade to a percentage mark (using
  * a rather mean marking scheme).
  */
public int letterToPerc(char letterGrade)
{
    int percGrade = -1;

    switch (letterGrade)
    {
        case 'F':
            percGrade = 35;
            break;

        case 'E':
             percGrade = 45;
            break;

        case 'D':
            percGrade = 55;
            break;

        case 'C':
            percGrade = 65;
            break;

        case 'B':
            percGrade = 75;
            break;
```

```
            case 'A':
                percGrade = 85;
                break;

            default:
                System.err.println("\nCould not covert mark into a %");
                break;
        }
        return percGrade;
    }
```

In addition to the `switch` statement, there are a few of new ideas in the code above.

First, you will recall that we used a new type of variable – the `String` to store text. Since this variable is itself a Java object, we can use any of the methods defined in the `String` class. Our example uses two of those methods. `compareTo()` performs an alphabetical comparison between the text typed in and the text 'A' and 'F'. The result of this method is 0 if the two strings match, −1 if the first is alphabetically before the second, and +1 if the first is alphabetically after the second. The second `String` method used is the `length()` method that counts the number of characters stored in the string (and therefore typed in by the user). Since we are only expecting a single letter grade, we can limit input to 1 character only.

Second, we use yet another new type of variable in the `letterToPerc()` method. The `char` primitive variable type is used to store single characters of text. Being a primitive (like `int` or `float`) as opposed to a class (like `String`) means that it does not have any of its own methods. We can use this character variable in the `switch` statement to find out which of the letters A–F have been typed in. Note that when comparing single characters, we use single quotes ' ' to delimit each case constant.

Finally, notice also that if Java cannot find a match between what has been typed in and the letters A–F, the resulting error message is printed using the method `System.err.println()`, rather than `System.out.println()`. The message is still printed on the screen, but as we shall see in Chapter 9, is useful for reporting error messages.

Text, ASCII and Unicode

When computers store text information, they require some form of coding that allows each text character to be represented as a binary number in the computer's memory. The traditional coding system to do this is known as ASCII (American, Standard Code for Information Interchange). In this system, 7 binary digits are used to store each character code. The resulting 128 character combinations are used to store 'Latin' characters A–Z in upper and lower case, 'Arabic' numerals 0–9 and other common symbols and punctuation marks (e.g. ., " + etc.).

Making Decisions 115

> A problem arises when trying to share textual information in a global environment where many non-Latin scripts are also used (e.g. Cyrillic, Hebrew, Hiragana etc.). *Unicode* provides a solution to the problem by using an internationally agreed 16-bit coding convention that contains not only all the characters that make up a given script, but also information on the type of script used. This provides an internationally translatable and extendible coding system appropriate for global communication. Java makes use of Unicode in its `String` class and `char` primitive (see the www.unicode.org for more details).

5.2.3 When to Use `switch`

There are no rigid rules, but if the following apply, `switch` is probably appropriate.

- When it makes things clearer — i.e. when more than about 2 or 3 `if-else` statements would be required if `switch` were not used.

- When actions are dependent on single instances of a condition (not ranges of numbers).

- When a default value is required.

5.3 OTHER WAYS OF MAKING DECISIONS

5.3.1 The Conditional Operator

The conditional operator is a very shorthand way of evaluating an expression and performing an action that is dependent on it. As the name suggests, it is an operator, and is used in the same way as + and – might be used in an expression.

The structure of the conditional operator is as follows:

```
(conditionalExpression) ? trueValue : falseValue
```

If the `conditionalExpression` is true, then the operator assigns `trueValue`. If it is false, the operator assigns `falseValue`. This rather cryptic definition makes more sense with an example:

```
max = (number1 > number2) ? number1 : number2;
```

The variable `max` will be set to `number1` if the condition (`number1>number2`) is true, otherwise it will be set to `number2`. The result of this is that `max` will hold the larger of the two numbers (or `number2` if they are both the same value).

The conditional operator can be a little confusing and can always be replaced with if-else. It is therefore advisable usually to avoid using it. However, programmers sometimes use it to perform simple evaluations such as the one above, so be prepared to recognise it when reading others' code.

5.4 GOOD DECISION-MAKING DESIGN

You will undoubtedly make considerable use of various decision-making constructions in your own Java programming. A part of good programming design involves thinking carefully about how decisions are made in your own methods. Frequent use of the if statement can sometimes indicate that there is scope for improvement in the efficiency of your code. Consider the following problem as an illustration of this process.

Suppose we wish to be able to display a floating point number on the screen to two decimal places. We can achieve this by some simple arithmetic and rounding:

```
float originalNumber = 65.7654321;
int scaledNumber   = (int)(originalNumber*100);
float roundedNumber = scaledNumber/100f;
System.out.println(roundedNumber);
```

By multiplying originalNumber by 100 before it is typecast into an integer, its last two decimal places are preserved. This method works in many circumstances, but float to integer typecasting will always round numbers *down*. So in the example above, the number 65.76 would be displayed when the correct number should be 65.77.

One way of overcoming this problem is to add an if statement that checks to see whether a number should be rounded up or down:

```
float originalNumber = 65.7654321;
int scaledNumber   = (int)(originalNumber*100);

if (originalNumber*100 - scaledNumber > 0.5)
    scaledNumber += 1;

float roundedNumber = scaledNumber/100f;
```

This would work and is a logical way to round an integer to the nearest whole number. However, a more efficient and simpler process can do without the if statement altogether:

```
float originalNumber = 65.7654321;
int roundedNumber = (int)((originalNumber+0.5)*100);
scaledNumber = roundedNumber/100f;
```

Making Decisions 117

> **Rounding Numbers**
>
> In practice, you are unlikely to use either of the processes above to round numbers, as Java contains several classes and methods that will do the job for you. In particular, Java contains the static method `Math.round()` that performs rounding of real numbers into integers and a class `DecimalFormat` that gives you much more control over the way real numbers are displayed.
>
> The next chapter will describe how these and other classes that come with the Java language can be explored using on-line documentation.

5.5 CASE STUDY: ADDING SPATIAL CLASSES TO GARDEN ANTS

You will recall from Section 3.5, that we developed a set of classes for modelling the behaviour of ants in a garden. In that section, we identified and coded classes that represented a garden, an ants' nest, ants and animals. In this section, we will add to that model by considering how we can use decision-making in Java to add spatial functionality to these classes.

At the heart of the spatial behaviour will be a new class `SpatialObject` from which all other spatial classes will be descended. The state of this class needs to contain sufficient information to place an object in space and define its boundary, while its behaviour must allow its location and extent to be compared with other spatial objects.

5.5.1 The `Footprint` Class

We have already seen in Section 4.7 that we can represent spatial features using the raster model of space. The `RasterMap` class we developed contained sufficient information to identify the location and outer boundaries of the raster, as well as the attributes of any cell within it. However, the raster model is not the only way of defining an object's spatial characteristics. We will develop an alternative 'vector' model more fully in Section 7.1, but for now we will start by using a simpler alternative whereby we model our ants, their nest and the garden with a rectangle with known dimensions and location. This rectangle is represented by the class `Footprint`.

By storing the boundaries of an object in its own class, we can later adapt it to model more complex spatial characteristics without having to change any other classes. For the moment, we will simply use the class for storing a 2D location, and a width and height. This information is sufficient for identifying any one of four rectangular boundaries of an object (see Figure 5.1). We will also store the type of footprint so that we can add new footprints in the future (for example, 3D boundaries, circular boundaries, etc.).

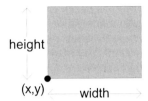

Figure 5.1 A 2D spatial footprint. The right-hand boundary is implicitly identified as x+width, the upper boundary as y+height.

The Java code for `Footprint` is shown below. It stores the geometry and type of boundary along with some simple accessor and mutator methods for extracting and changing these characteristics.

```
//    ****************************************************************
/** Class for defining the spatial footprint of an object.
  * Using this class to wrap all public communication of locations
  * and boundaries ensures maximum flexibility. Current model
  * assumes 2d Cartesian referencing, but can be adapted in future.
  * @author Jo Wood
  * @version 1.2, 6th September, 2001.
  */
//    ****************************************************************

public class Footprint
{
    // ---------------- Object variables --------------------

    private float  x,y,                // Point location/origin.
                   width,height;       // Width and height of MER.
    private int    type;               // Type of spatial footprint.


    // ------------------ Constructors ---------------------

    /** Creates a point footprint with given x,y location.
      * @param x X coordinate of footprint's location.
      * @param y Y coordinate of footprint's location.
      */
    public Footprint(float x, float y)
    {
        this.x = x;
        this.y = y;
        width  = 0;
        height = 0;
        type   = SpatialModel.POINT;
    }

    /** Creates 2d areal footprint with given origin, dimensions.
      * @param x X coordinate of footprint's origin.
      * @param y Y coordinate of footprint's origin.
      * @param width Width of object (in x direction).
      * @param height Height of object (in y direction).
      */
```

```java
    public Footprint(float x, float y, float width, float height)
    {
        this.x      = x;
        this.y      = y;
        this.width  = width;
        this.height = height;
        type        = SpatialModel.AREA;
    }

    // -------------- Accessor/mutator methods ---------------

    /** Reports the x origin of the footprint.
      * @return xOrigin Cartesian x coordinate of footprint.
      */
    public float getXOrigin()
    {
        return x;
    }

    /** Sets the x origin of the footprint.
      * @param xOrigin New x origin of the footprint.
      */
    public void setXOrigin(float xOrigin)
    {
        this.x = xOrigin;
    }

    /** Reports the y origin of the footprint.
      * @return yOrigin Cartesian y coordinate of footprint.
      */
    public float getYOrigin()
    {
        return y;
    }

    /** Sets the y origin of the footprint.
      * @param yOrigin New y origin of the footprint.
      */
    public void setYOrigin(float yOrigin)
    {
        this.y = yOrigin;
    }

    /** Reports width of footprint's Minimum Enclosing Rectangle.
      * @return width MER's width (in the x direction).
      */
    public float getMERWidth()
    {
        return width;
    }

    /** Sets width of footprint's Minimum Enclosing Rectangle.
      * @param width New width of the MER (in the x direction).
      */
    public void setMERWidth(float width)
    {
        this.width = width;
    }
```

```java
    /** Reports height of footprint's Minimum Enclosing Rectangle.
      * @return height MER's height (in the y direction).
      */
    public float getMERHeight()
    {
        return height;
    }

    /** Sets width of footprint's Minimum Enclosing Rectangle.
      * @param width New width of the MER (in the x direction).
      */
    public void setMERHeight(float height)
    {
        this.height  =height;
    }

    /** Reports the type of spatial footprint.
      * @return Type of spatial footprint.
      */
    public int getType()
    {
        return type;
    }

    /** Sets the type of spatial footprint.
      * @param type Type of spatial footprint.
      */
    public void setType(int type)
    {
        this.type = type;
    }

    // ------------------ Overridden methods --------------------

    /** Reports the details of this footprint.
      * @return Summary of this footprint.
      */
    public String toString()
    {
        return new String("Footprint with corners ("+x+","+y+
                          ") and ("+(x+width)+","+(y+height)+")");
    }
}
```

The final method `toString()` reports a textual summary of the object. All objects contain this method implicitly (inherited from the Java class `Object`) and is called automatically whenever an object is printed on-screen using `System.out.println()`. By overriding the default method, we can provide a more useful and informative message if ever `Footprint` is displayed in this way.

5.5.2 The `SpatialModel` Interface

Both the raster model of space, our simple footprint and the 'vector' model of space we will develop in Chapter 7 all have several characteristics in common. They can all be located in space with a set of coordinates, they all have an outer boundary and the raster and vector models can all contain one or more attributes at a given location. We might also wish to perform simple spatial comparisons with both models (for example, does spatial object *a* overlap with spatial model *b*?). It makes sense to identify this common state and behaviour and place it in its own class. That way, we can reduce repetition and increase the reusability of our classes. As we have seen in Section 3.1, this can be achieved using *inheritance generalisation*. When we do this, we have a choice of either creating a class or an interface containing the common state and behaviour. In this case, it makes sense to create an interface since the way in which we implement the behaviour may well depend on the type of spatial model used.

The code below shows the `SpatialModel` interface that contains sufficient information and method definitions to apply to a range of possible spatial models.

```
//   ***************************************************************
/** Identifies the state and behaviour of all spatial models.
  * @author Jo Wood.
  * @version 1.2, 16th September, 2001
  */
//   ***************************************************************

public interface SpatialModel
{
    // ------------------- Class variables ---------------------

    // Model type constants.
            /** Indentifies the spatial model is unknown. */
    public static final int UNKNOWN_MODEL = 100;
            /** Indentifies the spatial model as a 2d raster. */
    public static final int RASTER_2D = 101;
            /** Indentifies the spatial model as a 2d vector. */
    public static final int VECTOR_2D = 102;
            /** Indentifies the spatial model as a 3d raster. */
    public static final int RASTER_3D = 103;
            /** Indentifies the spatial model as a 3d vector. */
    public static final int VECTOR_3D = 104;

    // Query constants.
            /** Indicates value of spatial model is undefined. */
    public static final float NO_VALUE = Float.MIN_VALUE;
            /** Indicates query is outside bounds of model. */
    public static final float OUT_OF_BOUNDS = Float.MAX_VALUE;

    // Spatial type constants.
            /** Identifies a point object. */
    public static final int POINT = 0;
            /** Identifies a linear object. */
    public static final int LINE = 1;
            /** Identifies an areal object. */
    public static final int AREA = 2;
            /** Identifies a volumetric object. */
    public static final int VOLUME = 3;

    // Spatial comparison constants.
            /** Identifies this object as being within another. */
    public static final int WITHIN = 1;
```

```
        /** Identifies this object as matching another. */
    public static final int MATCHES = 2;
        /** Identifies this object as overlapping another. */
    public static final int OVERLAPS = 3;
        /** Identifies this object as enclosing another. */
    public static final int ENCLOSES = 4;
        /** Identifies object as being adjacent to another. */
    public static final int ADJACENT = 5;
        /** Identifies object as being separate from another. */
    public static final int SEPARATE = 6;


    // ------------------ Interface Methods --------------------

    /** Reports the type of spatial model.
      * @return Type of spatial model (RASTER_2D, VECTOR_3D etc).
      */
    public int getType();

    /** Reports the attribute of the model at the given location.
      * @param location Location to query.
      * @return Attribute at location or NO_VALUE if none defined.
      */
    public float getAttribute(Footprint location);

    /** Reports the outer boundaries of the object.
      * @return Outer boundary of the spatial object.
      */
    public Footprint getBounds();

    /** Sets the outer boundaries of the object.
      * @param bounds New outer boundary of the object.
      */
    public void setBounds(Footprint bounds);

    /** Performs spatial comparison between this object and another.
      * @param other Other spatial object to compare with this one.
      * @return Type of relationship (OVERLAPS, SEPARATE etc).
      */
    public int compare(SpatialModel other);
}
```

There is one new Java concept introduced by this interface, namely a new type of state – the *constant*. This uses the Java keywords `static final`. If a variable is declared `final` it means that its value cannot be changed once initialised. If it is declared `static` it means that the variable is stored as part of the *class*, not as part of any *object* created from it. This rather subtle distinction means that we can find out the value of a static variable without creating an object out of the class. Consequently, it would be perfectly valid for an object to include the line

`int relation = SpatialModel.SEPARATE;`

Note that we have used the *class* name (`SpatialModel`), not an object, to refer to the variable. For this reason, we shall refer to such variables as *class variables* in order to distinguish them from object or instance variables.

Making Decisions

Static constants are useful when creating *identifiers* or *symbols* that make reading code simpler. So, in our example, we have created six such identifiers, each symbolising a different type of spatial relation. By convention, such constants are given all upper-case letters. This helps us distinguish them from object variables.

Our interface also contains other useful constants for identifying possible types of spatial models (for example, a 3D raster) that we might use in the future. We can also add new constants at a later date without affecting existing classes.

The behaviour of the interface allows us to identify and change the type, boundary and attributes of the spatial model as well as perform some spatial comparison with another spatial model. Note also that all incoming or outgoing messages that refer to location or boundaries use the Footprint class. This ensures maximum flexibility if we wish to increase the complexity of our Footprint representation at a later date.

Figure 5.2 illustrates how this new interface can be integrated with the existing classes in our ants model.

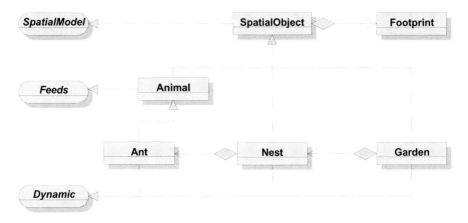

Figure 5.2 Garden ants class diagram incorporating the SpatialModel interface.

5.5.3 Comparing Spatial Objects

Our final task at this stage is to develop the SpatialObject class that acts as a parent for all other classes that have spatial characteristics (Animal, Nest and Garden). Its implementation is relatively straightforward, consisting of the accessor and mutator methods identified in the SpatialModel interface. The one exception is the method for performing spatial comparisons, which we will investigate here using the concepts of decision-making covered in this chapter.

Remember, we have kept the design of our spatial objects as simple rectangles with a known location, width and height. Now imagine we had two of these rectangular objects both of which are located somewhere on a plane. The relationship between the two will depend on their relative locations and size.

If we ignore the order in which we specify our rectangles (i.e. 'on top of' or 'below' have no meaning) as well as any sense of direction (i.e. to the 'left of' is no different to the 'right of'), there are six possible spatial relationships between two rectangles. These are illustrated in Figure 5.3.

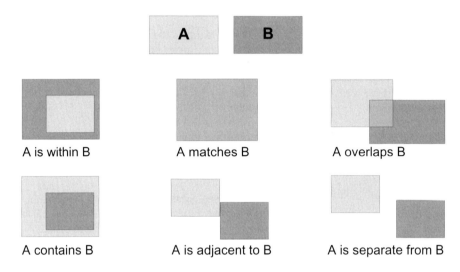

Figure 5.3 Six topologic relationships between two rectangles.

So, how do we identify these relationships based on the coordinates of the two rectangles? The answer is going to rely on several `if` statements that compare the four coordinates.

Before reading on, try to work out the 'if' rules you might use to identify these relationships. Remember, each rectangle is identified by and (x, y) location and (width, height) dimensions. The challenge is to construct a series of identifying rules with as few `if` statements as possible.

The code below shows one possible implementation of a `SpatialObject` class that identifies five of these relationships with four `if` statements.

```
//      ***********************************************************
/** Class for defining spatial objects. All such spatial
 *  objects have some geographical footprint and can compare
 *  themselves with other geographical footprints.
 *  @author Jo Wood
 *  @version 1.2, 16th September, 2001.
 */
//      ***********************************************************
```

```java
public class SpatialObject implements SpatialModel
{
    // ------------------ Object variables -------------------

    private Footprint footprint;        // Geographical footprint.

    // --------------------- Constructor ---------------------

    /** Creates a spatial object with the given spatial footprint.
      * @param footprint Spatial footprint of this object.
      */
    public SpatialObject(Footprint footprint)
    {
        this.footprint = footprint;
    }


    // ------------------------ Methods ----------------------

    /** Makes a spatial comparison between this object and another.
      * Assumes that both objects are rectangular.
      * @param other Spatial model with which to make a comparison.
      * @return One of five topologic relations (WITHIN, MATCHES,
      * OVERLAPS, ENCLOSES, or SEPARATE).
      */
    public int compare(SpatialModel other)
    {
        Footprint otherFp = other.getBounds();

        // Store Minimum enclosing rectangles of both footprints.
        float minX1 = footprint.getXOrigin(),
        minY1 = footprint.getYOrigin(),
        maxX1 = footprint.getXOrigin()+footprint.getMERWidth(),
        maxY1 = footprint.getYOrigin()+footprint.getMERHeight(),
        minX2 = otherFp.getXOrigin(),
        minY2 = otherFp.getYOrigin(),
        maxX2 = otherFp.getXOrigin()+otherFp.getMERWidth(),
        maxY2 = otherFp.getYOrigin()+otherFp.getMERHeight();

        // Compare the two rectangles.

        if ((maxX1 < minX2) || (minX1 > maxX2) ||
            (maxY1 < minY2) || (minY1 > maxY2))
            return SEPARATE;

        if ((minX1 > minX2) && (maxX1 < maxX2) &&
            (minY1 > minY2) && (maxY1 < maxY2))
            return WITHIN;

        if ((minX1 < minX2) && (maxX1 > maxX2) &&
            (minY1 < minY2) && (maxY1 > maxY2))
            return ENCLOSES;

        if ((minX1 == minX2) && (maxX1 == maxX2) &&
            (minY1 == minY2) && (maxY1 == maxY2))
            return MATCHES;
        return OVERLAPS;
    }
```

```java
    /** Moves spatial object relatively by the given coordinates.
     * @param xOffset Offset in the x direction.
     * @param yOffset Offset in the y direction.
     */
    public void move(float xOffset, float yOffset)
    {
        footprint.setXOrigin(footprint.getXOrigin()+xOffset);
        footprint.setYOrigin(footprint.getYOrigin()+yOffset);
    }

    /** Reports the details of this spatial object.
     * @return Summary of this spatial object.
     */
    public String toString()
    {
        return footprint.toString();
    }

    // --------------- Accessor/mutator variables ----------------

    /** Reports the type of spatial object.
     * @return Type of spatial object (e.g. LINE, AREA etc.).
     */
    public int getType()
    {
        return footprint.getType();
    }


   /** Reports the attribute of the object at the given location.
     * @param location Location to query.
     * @return Attribute at location or NO_VALUE if none defined.
     */
    public float getAttribute(Footprint location)
    {
        // Currently, object has no attribute value.
        return SpatialModel.NO_VALUE;
    }

    /** Sets the bounding rectangle of this object.
     * @param footprint Spatial footprint to assign to object.
     */
    public void setBounds(Footprint footprint)
    {
         this.footprint = footprint;
    }

    /** Returns the 2d bounding rectangle of the object.
     * @return Minimum enclosing rectangle of object.
     */
    public Footprint getBounds()
    {
        return footprint;
    }
}
```

The method `compare()` implements the spatial decision-making that determines which of five possible relationships might exist between two rectangles. An incoming message is passed to the method containing another object that implements `SpatialModel`. Because the class contains accessor methods for all four rectangle coordinates, we can compare the edges of both in order to identify their relationship.

We can translate the first `if` statement as reading *if the right-hand edge of one rectangle is to the left of the other, or the upper edge of one rectangle is below the lower edge of the other, the two must be separate.* We can perform similar tests for three of the other possible relationships.

Efficient decision-making is determined in part by the *order* in which decisions are made. So, once we have excluded separate and containing relations, we know that at least one edge of each rectangle must intersect. Note also that the most complicated relation to determine is probably the 'overlaps' one, but we can simplify this by testing for all others first. If none of the other conditions are met, we can conclude that the rectangles overlap by default.

Topologic Relations Between Objects

The categorisation of six different relationships between two areas described above is a simplification of the much larger set of possible topologic relationships between an arbitrary pair of spatial objects.

Egenhofer and Franzosa (1991) suggested a more thorough and formal approach based on a '9-intersection model'. They suggested that any bounded object consists of an *interior*, *exterior* and a *boundary*. Thus the topologic relationship between two such objects can give rise to nine possible intersection types and a maximum of 2^9 possible types of topologic relationship.

- Separate
- External adjacency
- Internal adjacency
- Intersection
- Enclosure
- Coincidence

Adapted from Molenaar (1998) p.110

> Molenaar (1998) considers how these relationships may be encoded for typical configurations of objects in 2D space. An example of the ten possible configurations of two areas is shown above. By providing a suitable coding of these relationships, it is possible to build a robust object-oriented framework for handling the topology of any combination of *n*-dimensional objects in *m*-dimensional space.

We can test our `compare()` method with a simple class that creates two `SpatialObjects` and performs a topologic comparison. Notice how the `switch` statement is used to 'decode' the spatial relationship identifiers.

```java
// ********************************************************
/** Program to create two rectangles and perform a
  * topologic comparison between them.
  * @Author   Jo Wood.
  * @Version  1.1, 17th September, 2001
  */
// ********************************************************

public class TestComparisons
{
    /** Creates two spatial objects and displays their topologic
      * relationship to one another.
      * @param args Command line arguments (ignored).
      */
    public static void main(String args[])
    {
        SpatialObject
                rect1 =  new SpatialObject(new Footprint(3,4,8,10)),
                rect2 =  new SpatialObject(new Footprint(0,0,5,5));

        System.out.println(rect1);

        int comparison = rect1.compare(rect2);

        switch (comparison)
        {
            case (SpatialModel.WITHIN):
              System.out.println("is within");
              break;

            case (SpatialModel.MATCHES):
              System.out.println("matches");
              break;

            case (SpatialModel.OVERLAPS):
              System.out.println("overlaps with");
              break;

            case (SpatialModel.ENCLOSES):
              System.out.println("encloses");
              break;

            case (SpatialModel.SEPARATE):
              System.out.println("is separate to ");
```

```
                break;
            default:
                System.out.println("has an unknown relation with ");
                break;
        }
        System.out.println(rect2);
    }
}
```

When run, this produces the following output:

```
Footprint with corners (3.0,4.0) and (11.0,14.0)
overlaps with
Footprint with corners (0.0,0.0) and (5.0,5.0)
```

That completes our initial design of the ants model and has provided us with a useful framework for the construction of further spatial models.

5.6 SUMMARY

We have seen in this chapter how we can further control the flow of program execution using various conditional statements. Using them effectively involves understanding the syntax of the if, else, switch and case statements (the easy bit), as well as the ability to implement decision-making in a clear, logical and efficient manner (the difficult bit).

Using the ants in the garden case study, we developed some spatial classes for storing the rectangular bounds of spatial objects. We used the ideas of programmatic decision-making to develop a method for identifying the types of topologic relationships that can exist between these objects.

By the end of this chapter and associated exercises on the web, you should be able to

- translate a problem into steps that involve conditional statements;
- know when to use each of the types of conditional statements in Java;
- know when it is appropriate to create object variables and when to create class variables; and
- understand how conditional statements can be used to identify the topological relationships between two spatial objects.

CHAPTER SIX

Sharing Classes

This chapter considers some of the ideas that will help you to share your classes with other programmers as well as introducing some further classes that come as part of the Java language. Effective Java programming relies on the fact that no single programmer will be able to create all the classes that are needed to perform many tasks, and that sharing classes and information about them is vital part of the programming process.

6.1 CONTROLLING VARIABLE SCOPE

When we create a variable or object in Java we have to make a decision as to how 'visible' that variable or object is to other bits of Java code. In doing so, we are deciding what aspects of our programs we are prepared to share with others, and what we wish to keep 'private'.

The general principle that we have been following so far has been to try and encapsulate code as much as possible. That is, restrict access to variables and methods to only those parts of code that need to use them. This philosophy of 'keep things private unless we have good reason to share' is a sound one in object-oriented programming. However, things are more complicated than keeping things either public or private. As programmers, we have control over the degree of privacy we wish to exercise of the state and behaviour of our classes. The extent to which a part of a program is visible to other parts is known as its *scope*. We can control the scope of a variable, method, or object by

- choosing where to declare it; and
- using the *modifier* `private`, `public` or `protected`.

We will consider the full range of program scope options from the most invisible (local) to most visible (global).

6.1.1 Local Variables

A number of Java statements involve enclosing some code inside braces { } (for example, `for` loops, `do-while` loops, `if` statements, etc.). It is possible to declare and use variables within these braces. In fact, we have already seen several examples of this, such as when we swapped the contents of two variables by creating a third temporary variable:

```
if (num1 > num2)
{
    int temp;
    temp = num1;
    num1 = num2;
    num2 = temp;
}
```

By creating the variable `temp` inside the braces, it is only visible to code that also lies within the same braces. As soon as the code in braces has been executed, the variable `temp` disappears (and memory is freed). This is the most local form of variable in that it only exists for a very small part of the program.

Such local variables should be used when (a) they are only required for a small part of the program; and (b) it makes the program clearer than declaring the variable at the top of the method in which it is used.

6.1.2 Method Variables

Many of the variables and objects we have created so far have been declared inside a method. By doing so, all parts of the method have access to these variables, but anything outside of the method does not know of their existence.

This is probably the most common form of encapsulation, and you should consider it the 'default' scope of variables. You should only create variables/objects with a greater scope than this if you have good reason to do so. For example, if more than one method in a class wishes to use the variable, but does not require message passing to do so.

6.1.3 Object Variables

It is sometimes desirable to create variables that several methods within a class might all use. For example, in our `SpatialObject` class in the previous chapter, the variable `footprint` was used by the constructor, accessor methods, `compare()`, `move()`, and `toString()`. Consequently, it was defined as an object variable near the top of the class declaration.

If you have good reason to grant more than one method access to a variable, it can be declared within the class definition, but before any methods. We have labelled such variables as object variables since they are useable by all code within an object. These are sometimes also known as *instance variables* as they are associated with an instance of the class (i.e. an object). Importantly, if several objects are created from the same class, each may store different values for each of the object variables (for example, two different `SpatialObjects` can store entirely different `footprints`).

As we have already seen, object variables come in three types: (1) variables that can only be used within an object (`private`); (2) variables that can be used both within and outside of an object (`public`); and (3) variables that can only be used within an object or any of its subclassed children (`protected`).

In fact, there is a fourth scope for object variables known as *package-wide* scope, which makes variables visible only to other classes in the same package. But more on this later on when we discuss the creation and use of packages of classes.

You should make object variables `private` by default unless you have good reason to make them accessible from outside the class in which they were created. Even if you wish other objects to change the value of object variables, it is better to use public accessor and mutator methods to control access to them (see Section 3.3.1 for a justification of this rule).

Similar rules apply to the scope of methods within a class. If you can guarantee that a given method will only ever be used by other methods in the same class, it should be declared private. More commonly for methods, we might expect them to be called from other classes – in such cases, they should be declared public.

6.1.4 Class Variables

Occasionally, we may wish to create a variable inside a class that is visible to all other objects, even if we have not made an object out of our class. Such variables (which we might consider the most 'global') are called *class variables* and are created in exactly the same way as object variables, but with the modifier `static` placed in front of its declaration.

An example of class variables created in this way are the variables `WITHIN`, `CONTAINS`, `OVERLAPPING`, etc. created as part of our `SpatialModel` interface. Because they are static, they are created (only once) at the same time as the class is defined, and before any objects are made out of it.

As we saw in the previous chapter, it will therefore be perfectly valid in Java to have a line such as

```
int status = SpatialModel.SEPARATE;
```

Note there is no reference to any objects here, simply the class name followed by the class variable.

Class variables should be used with caution as they are not very object-oriented. Their main use is in defining useful constants such as the ones in `SpatialModel`. Methods may also be declared static if they do not rely on any object variables inside a class definition. Again, their use is not very object-oriented, but have a limited application when defining 'helper methods' such as the maths functions `Math.cos()`, `Math.sin()`, `Math.round()` etc.

To consolidate, consider the following class that creates a number of variables and methods each with a different program scope.

```java
// ************************************************************
/** Class to demonstrate the scope of variables (but actually
 * does nothing very useful).
 * @author  Jo Wood.
 * @version 1.1, 6th June, 2001.
 */
// ************************************************************

public class Scope
{
    // ------------ Class and object variables --------------

    public static int classVar;

    public float     publicObjectVar;
    private double   privateObjectVar;
    protected int    protectedObjectVar;
    int              packageObjectVar;


    // -------------------- Methods ---------------------

    /** A dummy method to show where local variables are created.
     */
    public void publicMethod()
    {
        int methodVar = 7;

        if (methodVar > 0)
        {
            float localVar = 1.0f;
        }
    }

     /** A static method can only use local or static variables.
       */
     public static void staticMethod()
     {
         System.out.println(classVar);
     }


    // ----------------- Private Methods ---------------

    /** A dummy private method to show that methods have a
     * scope too.
     */
    private void privateMethod()
    {
        // Only other methods inside this class can call
        // this method.
    }
}
```

6.2 DOCUMENTING JAVA CODE

Because the developers of Java realised that many programmers would be likely to share and reuse classes that come with Java, they ensured that all the Java classes, methods and fields were well documented. A complete set of program documentation for the language is known as an API (Application Programming Interface) and can be found either on Sun's Java website (www.javasoft.com) or in the `docs/api` folder of your own Java installation (if you installed the documentation files).

The Java developers also realised that programmers may wish to share classes they had written themselves, so they created a mechanism by which programmers could produce their own documentation quickly and easily.

The most useful aspects of a class' documentation are a description of what the class does; a list of the public methods it contains; and a description of any messages that can be sent to or from public methods within the class. You have been documenting these elements already when you add comments to your programs, so Java allows you to create a series of HTML documents based on these comments automatically.

The program to do this is known as `javadoc` and can be called either from the command line in (for example using DOS window or Unix shell), or through your IDE such as BlueJ. To use `javadoc` from the command line you need to open a DOS box or Unix console and type a command with the following structure:

```
javadoc Java_file(s) [-author] [-version] [-d directory]
```

For example,

```
javadoc *.java -author -version
```

This would create a set of documents in HTML format out of all the Java files in the current directory including within the documents, the name of the author and version number of each class.

So, how does `javadoc` know which comments to include in the documentation? To include a comment with the document, you must use a variation of the multi-line comment which contains two opening `*`s rather than just the one. For example,

```
/** This comment would be included in the documentation
 */
// But this line would be ignored by javadoc.
```

Because we may have encapsulated some details inside our classes, we may not wish it all to be documented, so by default `javadoc` will only create documents for non-private classes, methods and fields.

All Classes	Package **Class** Tree Deprecated Index Help
	PREV CLASS NEXT CLASS FRAMES NO FRAMES All Classes
Flower	SUMMARY: NESTED | FIELD | CONSTR | METHOD DETAIL: FIELD | CONSTR | METHOD
FlowerArea	
GrassArea	## Class Plant
Park	`java.lang.Object`
Path	|
Plant	+--**Plant**
PlayArea	
RunPark	**Direct Known Subclasses:**
Tree	Flower, Tree
	public class **Plant**
	extends java.lang.Object
	Represents any plant. All plants have at least an age, a height and a location.
	Version:
	1.1, 27th August, 2001
	Author:
	Jo Wood
	### Field Summary
	`protected int` **`age`**
	Age of plant.
	`protected float` **`height`**
	Height of plant.
	`protected int` **`locationX`**
	x-coordinate of plant location.
	`protected int` **`locationY`**
	y-coordinate of plant location.
	### Constructor Summary
	`Plant``()`
	Initialises the height, age and location of the plant.
	### Method Summary
	`void` **`displayDetails`**`()`
	Displays details of the plant on the screen.
	`void` **`grow`**`()`
	Doubles the height of the plant and ages it by a year.
	Methods inherited from class java.lang.Object
	`clone, equals, finalize, getClass, hashCode, notify, notifyAll, toString, wait, wait, wait`

Figure 6.1 Part of the `javadoc` output for the `Park` classes.

If you wish your name (as author) and version number of each class to appear in the documentation, you must add the @author and @version lines to the header of your programs. You should also describe any incoming messages to a method with the @param tag and outgoing messages with the @return tag.

As an example, Figure 6.1 shows part of the javadoc pages generated from the Park class created in earlier chapters. The document contains an index of all methods and constructors inside the class as well as more detailed information on each further down the document. If you have several classes to document (as in the example), javadoc will create an alphabetical index of all the classes, methods and fields, and cross-reference them if necessary. This can save a large amount of time and effort and makes it much easier for others to use and understand your programs.

6.3 JAVA PACKAGES

Most Java programs of at least moderate complexity are likely to use and contain many classes from which objects are created. In order to manage the many hundreds of classes the programmer is likely to be exposed to, the Java language allows classes to be grouped together in *packages* – collections of classes with similar themes. This section considers some of the packages built into the Java language as well as how to organise new classes into their own packages.

6.3.1 Exploring Java's Packages

The standard edition of Java 1.4 contains over 2700 different classes arranged into over 130 packages. These classes range widely from classes to connect to relational databases, through cryptographic encoding, to the manipulation of sound samples. So, how are we to find out the classes available to us, and how do we use the methods contained within them? The answer to both questions involves reading the substantial API documentation (stored as web pages) that comes with the language. This API, which should be provided with your Java installation, or can be viewed directly from Sun's own website, is created using javadoc and will have the same appearance as your own program documentation.

Packages are arranged hierarchically using a lower-case naming convention, separating each 'level' of the hierarchy with a dot (.). So for example, the following are all Java package names:

```
java.awt
java.awt.image
java.awt.image.renderable
```

As we have already seen, the java.awt package contains many of the classes necessary to create a GUI using the Abstract Windowing Toolkit. Within that package is a further package containing classes specifically related to handling images. Within that, is one further package containing classes with responsibility for rendering images.

The packages you are likely to use in your own programs can be grouped into four types as shown in Table 6.1.

Table 6.1 Four types of Java packages

Package type	Example class	Explanation
java.lang	String	Anything in this package is considered part of the heart of the language is readily available to the programmer without explicit importing.
java.-	java.awt.Graphics	These 'core' classes should be available in any implementation of the language on any platform.
javax.-	javax.swing.JFrame	The 'extension' classes are provided by the vendor of the language, but may not be available in all implementations.
-.-	org.w3d.dom.Document	Pacakges that do not start with java. or javax. are provided by 'third parties' and might include those you develop yourself.

To use any of the classes in packages other than java.lang, you need to *import* them into your own programs. This is done with the import keyword followed by the package name and list of classes to import. For example, to import all of the classes for creating a graphical user interface, you might include the lines

```
import java.awt.*;
import javax.swing.*;
```

above your class definition. Including these lines in your program is the equivalent of copying all the class definitions inside the AWT and Swing packages over to your current directory. Note that there is little overhead in doing this – only classes that you actually use will be compiled when you compile your Java classes. Note that using the * as a 'wildcard' like this will not import sub-packages; these have to be imported explicitly. For example,

```
import java.awt.*;
import java.awt.image.*;
```

In this way, you can take advantage of pre-written classes, which in many cases will save you much programming work yourself. So, before you consider writing a class to perform a particular task, it is always worth consulting the Java documentation to see if there is already an existing one that does the job for you.

Of the many hundreds of packages available to the Java programmer, there are some that you will almost certainly come across as you develop your programming skills. These are shown in Table 6.2.

Table 6.2 Some commonly used Java packages

Package	Useful classes
java.awt	Basic graphical user interface classes. Even if you use Swing this package contains useful classes such as Graphics, layout managers, and containers.
java.awt.event	Required for graphical event handling such as button presses, menu selection etc. (see Chapter 8).
javax.swing	The main GUI classes, mostly prefixed with a J (e.g. JFrame, JButton etc.) are stored in this package.
java.util	A range of utilities in this package, but particularly useful are the various 'collection' packages such as Vector, HashMap etc. (see Chapter 7) and the String tokenizer (see Chapter 9).
java.io	Input/output classes. Useful when handling files (see Chapter 9)
java.applet	Contains the Applet class used for creating web-embedded Java programs (see Chapter 10).

6.3.2 Creating Packages

As you begin to create more complex programs in Java, it increasingly makes sense to organise your own programs into their own packages. This is particularly the case once you start reusing classes you have written in a variety of other programs.

To place one of your own classes in a package, simply name the package in which the class is to sit at the top of the class definition. Packages should be named with lower-case letters only (i.e. no intercapping). As you may have seen from the existing Java documentation, a package can contain sub-packages within it. This hierarchical organisation helps keep large collections of classes manageable.

So, for example, the classes that I have written all sit somewhere inside the package called jwo (my own username and e-mail address). Within that package, it is possible to create further sub-packages, such as those for this book (jwo.jpss). Within that, there I might create a further package to store the park classes developed in previous chapters (jwo.jpss.park).

If arranged in this way, each of the park classes will then contain the line

```
package jwo.jpss.park;
```

as the first line. This indicates to Java that the class sits inside a package called park that sits inside another called jpss, which in turn sits inside one called jwo. Using this form of package naming prevents any naming conflicts with other classes you have written, or indeed those that form part of the Java language.

Whenever you create a package or sub-packages, the file representing each class must sit in a directory or sub-directory that corresponds to the package name. So, for example, the file storing the `Tree` class developed as part of the park model would be located at `jwo/jpss/park/Tree.java` somewhere on a file system readable by Java.

> ### Package Namespaces
>
> The term 'namespace' is sometimes given to indicate the rules for naming shared resources that avoid two developers independently coming up with the same name for a resource. This is particularly important for a language like Java where classes are designed for sharing over the internet.
>
> For packages that are publicly distributed or used commercially, the Java developers recommend the following naming convention for avoiding naming conflicts.
>
> Companies or organisations producing sharable classes should use their reversed internet domain name as the start point for all packages. So, for example, the World Wide Web Consortium (W3C) use `org.w3c`, the Object Management Group (OMG) use `org.omg` as their start point.
>
> Within any organisation, some agreed convention should be adopted to avoid namespace 'collisions'.
>
> User-created packages should *never* start with `java.` or `javax.` as this would conflict with the official licensed classes that make up the Java language.
>
> Packages created by individuals who do not belong to organisations with a reserved domain name should use a similar logical naming convention that is unlikely to lead to namespace collisions.

To use classes that have been created inside a class, even if they are ones you created yourself, you must first import the relevant package using the `import` statement. So, for example, we might create a `RunPark` class that contains the following line at the top of the file:

`import jwo.jpss.park.*;  // Imports all the park classes`

Creating packages can seem like a bit of an inconvenience at first, but the effort soon pays of as you start to develop 'libraries' of useful classes that can quickly be incorporated into new programs. For example, so far in this book, we have developed three groups of classes, all of which may be reused at some point in the future. Figure 6.2 shows a possible arrangement of these classes into packages.

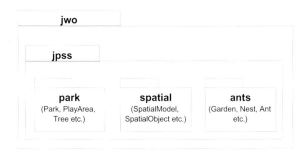

Figure 6.2 Three reusable packages inside the `jwo.jpss` package.

6.3.3 Archiving Packages and Classpaths

A package of classes is likely to consist of many files each representing a single class. If you wish to distribute a package of classes to others, the transfer of these files can be cumbersome, and if distributed over the internet, time-consuming to download.

To avoid this problem, the Java toolkit comes with an archiving and compression tool called `jar` (Java ARchiver). This tool will gather all the files in a folder into a single .jar file and compress them in the process. The compression is compatible with the widely used ZIP compression used on PC platforms.

To archive a collection of Java files, a command line shell (DOS box or Unix shell) must be opened and the command with the following form given:

`jar -cvf archiveName javaFileNames`

where `archiveName` is the name of the JAR file to hold the archive and `javaFileNames` lists the files to be contained within the archive. This list can include the * wildcard character and can recursively search sub-folders.

For example, to archive all the classes in the `jwo` package, the following command could be given in the folder containing the `jwo` sub-folder:

`jar -cvf jwo.jar *`

This command would archive all files including source code, sub-folders and documentation contained within the directory from which the command was given. To be more selective, we might choose to archive only the compiled bytecode files within the `park` sub-package (note that 'forward slashes' can be used to differentiate directories regardless of operating system):

`jar -cvf jwo.jar jwo/jpss/park/*.class`

Classpaths

If you wish to run a program that is dependent on a variety of packages, class files or JAR files we need some way of informing the Java Virtual Machine where these files are located. Up until now, we have assumed that either the IDE knows where to look or that the `java` command was issued in the same folder as the bytecode to be interpreted.

As the number of packages you use increases, it is unlikely that all packages will be located in the same folder on your computer. In such cases, it is possible to give the Java interpreter a list of folders in which to look when assembling bytecode from a range of locations. This can be done from the command line by using the -classpath option. For example,

```
java -classpath c:\java\myClassFiles RunPark
java -classpath c:\java\jwo.jar RunAnts
java -classpath c:\java\jwo.jar;d:\apps\landserf.jar LandSerf
```

If more than one location is to be searched for, each should be separated by a semicolon on windows platforms or a colon on Unix platforms. For more details on archiving packages, see Sommerer (2001).

6.4 CASE STUDY: CREATING A GRAPHICAL ANTS MODEL

If you recall from Section 5.5, we created an object-oriented model to represent a nest of ants in a garden, but as yet we have no way of observing the model. In this section, we will create a simple graphical interface that displays an ant as it wanders around the garden.

As we have seen, all Java programs that create graphical output ultimately need to be contained in a window of some kind. So far, we have created windows in Java using the `JFrame` class that belongs to the Swing package. We will continue to do this in this case study, but you should bear in mind that it is possible to create other types of containers such as applets (see Chapter 10) and AWT `Frames`. In order to keep our programming as flexible as possible, we will design a graphical implementation that could work with any type of container. We will do this by introducing a new programming concept – that of *delegation*.

6.4.1 Keeping Graphics Object-Oriented

Remember that one of the goals of object-oriented programming is to create classes that are as independent as possible from one another. These classes should communicate where necessary by sending and receiving messages. We will bear this in mind when we consider how to implement the drawing code in our ant model.

Sharing Classes 143

What to draw?
Four elements need to be rendered by our classes, namely the containing graphics window, the garden, nest and the ants. Good object-oriented design would suggest that each of these classes should take responsibility for drawing their respective component of the model. In other words, the Ant class should know how to draw an ant, the Nest class should know how to draw a nest etc. In that way, we can continue to develop each class independently of the others without compromising the overall class design.

Where to draw?
The location of each graphical component obviously depends on what is being drawn. We would like the garden to take up most of the container window, the nest to appear somewhere in the middle of the garden and the ants to move about the garden. Since Garden, Nest and Ant are all SpatialObjects, keeping track of their location should not be too much of a problem. However, drawing these on the screen will involve some kind of *spatial transformation* that depends on the location and size of the container window. Ultimately therefore, all graphics rendering should be controlled via a class that is aware of the size and location of the container window.

When to draw?
Some graphical components like the container window and the garden need to be drawn when the program first starts, but then are unlikely to change unless the window itself is moved or resized. However, the ants that will be wandering all over the garden need to be repeatedly redrawn every time they move. We therefore need a mechanism by which an ant (or possibly a nest or garden) can tell the application window '*redraw me because I have changed since I was last drawn*'.

6.4.2 What to Draw

Object-oriented design suggests that responsibility for deciding what to draw as part of our GUI should be devolved to each relevant class. These are summarised in Table 6.3.

Table 6.3 Drawing procedure for the ant simulation

Class	What to draw
AntApplication	Container window with simple border.
Garden	Large green rectangular area.
Nest	Circular nest symbol at the centre of the garden.
Ant	Small black oval within the garden area.

The main container class that provides the start point for our application simply extends the JFrame class as we have done when creating other GUIs. The spatial classes (Footprint, SpatialObject and SpatialModel) have all been moved into their own package (jwo.jpss.spatial), as have the remaining ant simulation classes (jwo.jpss.ants).

```java
package jwo.jpss.ants;        // Part of the ant simulation package.

import javax.swing.*;         // Required for graphical objects.
import java.awt.*;
import jwo.jpss.spatial.*;    // For spatial classes.

// ****************************************************************
/** A top level application window for displaying ant activity.
  * @author   Jo Wood.
  * @version  1.3, 18th September, 2001
  */
// ****************************************************************

public class AntApplication extends JFrame
{
    // ----------------- Starter method --------------------

    /** Creates the graphical window within which ants can be
      * observed.
      * @param args Command line parameters (ignored).
      */
    public static void main(String args[])
    {
        new AntApplication("Ants in the garden...");
    }


    // ---------------- Object variables ------------------

    private Garden garden;        // Garden containing ants.

    // ------------------ Constructor ---------------------

    /** Creates a top level application window with a given title.
      * @param title Title to associate with window.
      */
    public AntApplication(String title)
    {
        super(title);
        setDefaultCloseOperation(WindowConstants.DISPOSE_ON_CLOSE);

        // Size window to fit around garden.
        GardenPanel gardenPanel = new GardenPanel();
        gardenPanel.setBorder(BorderFactory.createEtchedBorder());
        Footprint fp = new Footprint(0,0,400,400);
        garden = new Garden(fp);
        gardenPanel.setPreferredSize(
                        new Dimension((int)fp.getMERWidth(),
                                      (int)fp.getMERHeight()));
        getContentPane().add(gardenPanel);
        pack();
        setVisible(true);

        garden.startEvolution();
    }
}
```

This class simply creates a new `Garden` object and a `GardenPanel` on which to display it (see Section 6.4.3 below), and adds the panel to the window along with a simple border. The final line of the constructor starts the ants off on their wanderings around the garden.

The remaining three drawable classes (`Garden`, `Nest` and `Ant`) all use the graphics context provided to them by the container class (`AntApplication`). Because all of them should exhibit the same drawable behaviour, it makes sense to create a new interface called `Drawable` containing one abstract method `paint()`. All three classes can then implement this interface (at this stage our drawing behaviour is fairly simple, so creating a `Drawable` interface might seem unnecessary, but as we shall see later on when we allow more sophisticated drawing behaviour, this interface helps us keep our programming tidy).

Drawing code for `Garden`

```
private Rectangle bounds;        // Boundary of the garden.

/** Creates a garden with ants' nest and given spatial footprint.
  * @param footprint Spatial footprint of garden.
  */
public Garden(Footprint footprint)
{
    super(footprint);
    antsNest = new Nest(new Footprint(190,190,20,20));
    Footprint fp = getBounds();
    bounds = new Rectangle((int)fp.getXOrigin(),
                           (int)fp.getYOrigin(),
                           (int)fp.getMERWidth(),
                           (int)fp.getMERHeight());
}

/** Draws the garden and all contained within it using the
  * given graphics context.
  * @param g Graphics context to draw to.
  */
public void paint(Graphics g)
{
    // Draw garden.
    g.setColor(new Color(200,220,200));
    g.fillRect(bounds.x, bounds.y, bounds.width, bounds.height);

    // Draw nest.
    antsNest.paint(g);
}
```

The `Garden` class creates a suitably sized area (the bounds of which are stored in the AWT class `Rectangle`) and places a nest in its centre. The `paint()` method sets a pale green background colour and then instructs the nest to draw itself.

Drawing code for Nest

```
private Rectangle bounds;        // Boundary of the nest.

/** Creates a nest with a given spatial footprint.
  * @param footprint Spatial footprint of the nest.
  */
public Nest(Footprint footprint)
{
    super(footprint);

    // Store drawable size of nest.
    bounds = new Rectangle((int)footprint.getXOrigin(),
                           (int)footprint.getYOrigin(),
                           (int)footprint.getMERWidth(),
                           (int)footprint.getMERHeight());

    // Add 1 ant with 1000 food units at the nest location.
    ants = new Ant(1000, new Footprint(footprint.getXOrigin(),
                                       footprint.getYOrigin(),
                                       8,5));
}

/** Draws the nest.
  * @param g Graphics context in which to draw the nest.
  */
public void paint(Graphics g)
{
    // Draw nest as a simple circle.
    g.setColor(new Color(155,155,100));
    g.fillOval(bounds.x, bounds.y, bounds.width, bounds.height);
    g.setColor(Color.black);
    g.drawOval(bounds.x, bounds.y, bounds.width, bounds.height);

    // Draw ants associated with this nest.
    ants.paint(g);
}
```

The drawing code for Nest is similar to that for the garden. In both cases, the size and location of the nest is unlikely to change, so they are calculated in the constructor. The Nest class creates a single ant with 1000 food units and instructs it to draw itself.

Drawing code for Ant

```
/** Draws the ant using the given graphics context.
  * @param g Graphics context to draw to.
  */
public void paint(Graphics g)
{
    Footprint fp = getBounds();

    if (getFoodLevel() > 500)
        g.fillOval((int)fp.getXOrigin(), (int)fp.getYOrigin(),
                   (int)fp.getMERWidth(),(int)fp.getMERHeight());
    else
```

```
            g.drawOval((int)fp.getXOrigin(), (int)fp.getYOrigin(),
                       (int)fp.getMERWidth(),(int)fp.getMERHeight());
}
```

Since the ant will move around the garden, it is necessary to recalculate its position every time the ant is drawn. The drawing code itself is simple, drawing a solid oval (using the `Graphics` class' method `g.drawOval()`) if the ant carries more than 500 units of food, or otherwise a hollow outline.

6.4.3 Where and When to Draw

Ultimately, all graphics are displayed in a container class (`AntApplication` in our example). This is the class that creates a `GardenPanel`, which sets up the graphics context that all `paint()` methods use. This class needs to know not only about the graphics context, but how big the container is and how to get the `Garden`, `Nest` and `Ants` to draw themselves. This raises a problem because we wish to design our classes to be able to draw in a range of possible containers. How can we guarantee that the container will know its own size and when to ask any of the drawable classes to redraw themselves? A sensible object-oriented answer is to create another interface that guarantees the necessary functionality. This interface requires two methods – one to check whether the container is large enough to draw one of its contained objects, the other to tell a contained component to redraw itself (because it has changed since last being drawn). For reasons that will become apparent later, we shall call this interface `GraphicsListener`.

```
package jwo.jpss.ants;      // Part of the ant simulation package.

import jwo.jpss.spatial.*;  // For spatial classes.
//      ***********************************************************
/** Interface for graphical components that need to display
 *  dynamic objects that get updated periodically.
 *  @author Jo Wood
 *  @version 1.3, 20th August, 2001.
 */
//      ***********************************************************
public interface GraphicsListener
{
    /** Update graphics.
      */
    public abstract void redrawGraphics();

    /** Checks that a spatial object can be drawn by the listener.
      * @param footprint Spatial object we wish to draw.
      * @return True if the spatial object can be drawn.
      */
    public abstract boolean canDraw(SpatialObject spatialObject);
}
```

Any container class that implements this interface must therefore contain methods that will issue a redrawing command when asked, and will check that the object being drawn is actually located within the bounds of the container. In our example, this is easily implemented by creating a `GardenPanel` that extends Swing's `JPanel` class.

```
/** Simple panel for displaying a dynamic garden.
  */
private class GardenPanel extends JPanel
                    implements GraphicsListener
{
    /** Draws the simulation on the panel.
      * @param g Graphics context within which to draw.
      */
    public void paintComponent(Graphics g)
    {
        garden.paint(g);
    }

    /** Checks whether given spatial object can be drawn on panel.
      * @param footprint Spatial object we wish to draw.
      * @return True if the spatial object can be drawn.
      */
    public boolean canDraw(SpatialObject spatialObject)
    {
        if (garden.compare(spatialObject) == SpatialModel.ENCLOSES)
            return true;
        else
            return false;
    }

    /** Redraws any graphics that need updating.
      */
    public void redrawGraphics()
    {
        Dimension d = getSize();
        paintImmediately(0,0,d.width,d.height);
    }
}
```

`GardenPanel` overrides `JPanel`'s `paintComponent()` method and requests the `Garden` to draw itself. It implements the `canDraw()` method by checking to see if any drawable object is contained within the `Garden`. If it is, it allows the object to be redrawn. This spatial comparison comes 'for free' since it is built into the `SpatialObject` class. By implementing the method in this way, we can guarantee that none of our ants will be able to escape from the `Garden`.

The `redrawGraphics()` method calls the method `paintImmediately()` inherited from `JComponent`. This forces Java to repaint the container, effectively calling the `paintComponent()` of `GardenPanel`. In this way, whenever an object such as an Ant asks to be redrawn, the `GardenPanel` will oblige without delay.

Our final task is to establish a connection between all the drawable classes and the `GraphicsListener`. We can achieve this by adopting a *delegation model of event handling*. We tell `AntApplication` to *listen* for ants that have moved and redraw them when necessary. This is done using the `redrawGraphics()` method of our `GraphicsListener` interface.

So, if our container class is doing the listening, how do we get the ants to broadcast the fact that they have moved and therefore need redrawing? The answer is to pass a reference to the `GraphicsListener` down to the ants, so they can call the `redrawGraphics()` method directly. In fact, we only need pass the `GraphicsListener` as far as the `Nest` class since it is this one that keeps track of the ants in our simulation.

We shall pass the `GraphicsListener` from one class to another using a method called `addGraphicsListener()` which is placed in both the `Garden` and `Nest` classes. `Nest` then checks to see whether any ants are alive and need redrawing before calling the `updateGraphics()` method of the `GraphicsListener`.

Delegation code for AntApplication

```
/** Creates a top level application window with a given title.
  * @param title Title to associate with window.
  */
public AntApplication(String title)
{
    // AntApplication constructor here.
    garden.addGraphicsListener(gardenPanel);
}
```

Delegation code for Garden

```
/** Adds a graphics listener to this object. Passes on the listener
  * to the ants' nest.
  * @param graphicsListener Component doing the drawing.
  */
public void addGraphicsListener(GraphicsListener graphicsListener)
{
    antsNest.addGraphicsListener(graphicsListener);
}
```

Delegation code for Nest

```
/** Adds a graphics listener to this object. Allows graphics to be
  * drawn by a GraphicsListener.
  * @param graphicsListener Component doing the drawing.
  */
public void addGraphicsListener(GraphicsListener graphicsListener)
{
    this.graphicsListener = graphicsListener;
```

```
        // Add a graphics listener to each ant.
        ants.addGraphicsListener(graphicsListener);
}
```

The Ant Class

```
package jwo.jpss.ants;       // Part of the ant simulation package.

import java.awt.*;           // For drawing the ant.
import jwo.jpss.spatial.*;   // For spatial classes.

// ***********************************************************
/** Class for representing and drawing Ants.
  * @author Jo Wood
  * @version 1.2, 23rd August, 2001.
  */
// ***********************************************************

public class Ant extends Animal implements Dynamic, Drawable
{
    // ------------------- Object variables ----------------

    private int    straightSteps, // Number of steps taken in a
                                  // single direction.
                   numSteps;      // Number of steps walked by ant.
    private float  xDir,yDir;     // Current direction taken by ant.

                                  // Component drawing the graphics.
    private GraphicsListener graphicsListener;


    // --------------------- Constructor -------------------

    /** Creates ant with given initial food level and footprint.
      * @param foodLevel Initial food level of the ant.
      * @param footprint Initial spatial footprint of the ant.
      */
    public Ant(int foodLevel, Footprint footprint)
    {
        super(foodLevel,footprint);

        straightSteps = 20;
        numSteps      = (int)(Math.random()*10);
    }


    // ----------------- Implemented methods ---------------

    /** Draws the ant using the given graphics context.
      * @param g Graphics context to draw to.
      */
    public void paint(Graphics g)
    {
        Footprint fp = getBounds();
```

```java
        if (getFoodLevel() > 500)
           g.fillOval((int)fp.getXOrigin(), (int)fp.getYOrigin(),
                      (int)fp.getMERWidth(),(int)fp.getMERHeight());
        else
           g.drawOval((int)fp.getXOrigin(), (int)fp.getYOrigin(),
                      (int)fp.getMERWidth(),(int)fp.getMERHeight());
    }

    /** Adds a graphics listener to this ant. Allows graphics to be
      * drawn by a GraphicsListener.
      * @param graphicsListener Component doing the drawing.
      */
    public void addGraphicsListener(GraphicsListener gListener)
    {
        this.graphicsListener = gListener;
    }

    /** Let the ant go about its business for one time unit.
      */
    public void evolve()
    {
        metabolise(1);            // Use up 1 food unit.

        if (isAlive())
        {
            // Change direction if any has walked sufficient steps.
            if (numSteps%straightSteps == 0)
            {
                xDir = (float)(Math.random()*2-1);
                yDir = (float)(Math.random()*2-1);
            }

            move(xDir,yDir);

            // Only move if within bounds.
            if (graphicsListener.canDraw(this))
            {
                numSteps++;
            }
            else        // Change direction if cannot move.
             {
                move(-xDir,-yDir);
                xDir = (float)(Math.random()*2-1);
                yDir = (float)(Math.random()*2-1);
            }
        }
    }
}
```

After all that delegation, we now have a working set of classes that can display dynamically, the movement of an ant around the garden. The overall class structure is shown in Figure 6.3 and a sample of graphical output in Figure 6.4.

152   *Java Programming for Spatial Sciences*

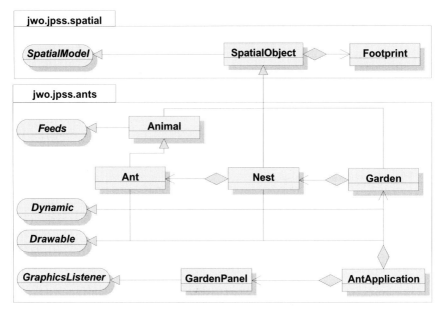

Figure 6.3 Class diagram showing drawable ant simulation.

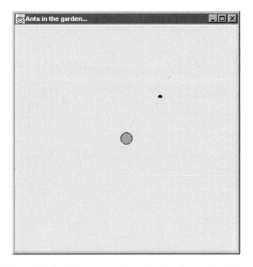

Figure 6.4 Graphical output from the simple ant simulation.

Sharing Classes 153

6.5 SUMMARY

We have seen how we can control the degree to which we make classes, methods and variables sharable. Once we appreciate that the classes we create may be used by other programmers, it becomes necessary to make use of those classes as simple as possible. Documenting code as we write it and creating understandable program documentation is a vital part of that sharing process. We have seen how the `javadoc` program can help in automatically compiling program documentation.

You should get into the habit of making frequent use of the existing Java API documentation when designing your own programs. It is likely that many of the programming tasks you wish to achieve will be considerably eased by using existing classes, either as part of the core Java language or as external packages written by others.

Arranging classes into packages helps organise increasingly complex programs that make use of many classes, perhaps written by many people. Packages should be named in such a way as to avoid namespace conflicts with those written by other programmers.

We have illustrated the creation and use of packages of sharable classes by linking previously developed spatial and ant modelling classes with the GUI classes that from the AWT and Swing packages.

By the end of this chapter and associated exercises on the web, you should be able to

- control the scope of a variable or method;
- know when to create variables and methods of different program scopes;
- create your own package of classes;
- import other programmers packages into your own programs;
- use comments and `javadoc` to provide useful documentation for your classes; and
- know where to look in order to find out about existing classes that come with Java.

CHAPTER SEVEN
Collecting Objects Together

Remembering that computer programming involves modelling aspects of the real world, we can recognise that in reality many items of information are often associated. For example:

- The letters that make up this sentence.
- The annual level of unemployment for the last 20 years.
- The coordinates that make up a polygon boundary.
- The cells that make up a raster model.

Java provides us with several mechanisms for grouping items of data. We have already used one of those – the array – when we considered how we might store raster images in Chapter 4. This chapter considers further applications of arrays and introduces the *dynamic collection* as an alternative and more flexible way of grouping objects together. In doing so, we can develop further models of spatial data.

7.1 VECTOR MODELLING OF SPATIAL OBJECTS

7.1.1 Using Arrays

As we saw in Chapter 4, an array is simply a list of variables that share the same name, but can be uniquely identified by their numerical index. We are not restricted in the type of variable that can be used in an array, which might vary from simple integers to complex objects.

Let us consider how we might store the coordinates that can be used to represent a line. It is common practice within GIS to represent the boundaries of things such as buildings, lakes, roads and cities, as groups of (x, y) coordinate pairs.

Figure 7.1 shows how we might model a river and two buildings using collections of coordinate pairs. Each pair represents a location, which is joined to the next pair of coordinates in an ordered sequence. The collection as a whole can be used to represent the river network and the building outlines.

This form of spatial modelling where we use coordinate pairs to represent the boundaries of things is known as the *vector* model of spatial information, and relies upon the ability to identify object *boundaries*.

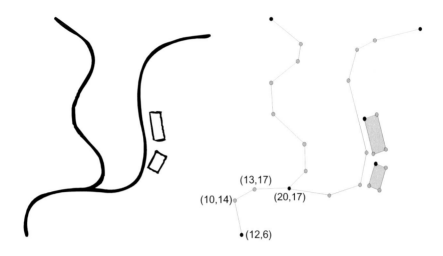

Figure 7.1 A vector representation of simple linear and areal features.

When we developed the spatial classes in order to model garden ants in Chapter 5, we created a Footprint class that can be used to store such coordinate pairs. By creating an array of Footprints, we have the basis for storing spatial boundaries. So, for example, to model a line with 10 coordinates we can create the following array of Footprint objects:

```
Footprint line[] = new Footprint[10];
```

It then becomes a relatively simple task to create as a (GIS) vector class based on our previously written SpatialObject class. This new class can be added to the jwo.jpss.spatial package (joining Footprint, SpatialModel and SpatialObject) created in the previous chapter.

```
package jwo.jpss.spatial; // Part of the spatial modelling package.

// ****************************************************************
/** Models a GIS vector object.
  * @author Jo Wood
  * @version 2.2, 3rd October, 2001
  */
// ****************************************************************
public class GISVector extends SpatialObject
{
    // -------------------- Object variables --------------------

    private Footprint[] coordinates;    // Vector geometry array.
    private int type;                   // Type of vector object.

    // --------------------- Constructor -----------------------
```

```
    /** Creates GIS vector object with the given x and y coords.
      * @param xCoords Array holding x-coordinates of object.
      * @param yCoords Array holding y-coordinates of object.
      * @param type Type of vector (POINT, LINE, AREA, VOLUME).
      */
    public GISVector(float[] xCoords, float yCoords[], int type)
    {
        // Initialise vector information and check integrity.
        int numCoords = xCoords.length;
        this.type = type;
        coordinates = new Footprint[numCoords];
        float north = -Float.MAX_VALUE;
        float south =  Float.MAX_VALUE;
        float east  = -Float.MAX_VALUE;
        float west  =  Float.MAX_VALUE;

        if (yCoords.length != numCoords)
            System.err.println(
                "Warning: Number of x and y coords do not match");

        if ((type==POINT) && (numCoords != 1))
            System.err.println(
               "Warning: Point does not contain 1 coordinate pair");

        if ((type==LINE) && (numCoords < 2))
            System.err.println(
                "Warning: Line contains less than 2 coordinates");

        if ((type==AREA) && (numCoords < 3))
            System.err.println(
                "Warning: Area contains less than 3 coordinates");

        if ((type==VOLUME) && (numCoords < 4))
            System.err.println(
               "Warning: Volume contains less than 4 coordinates");

        // Store the coordinates and calculate bounds of object.
        for (int i=0; i<numCoords; i++)
        {
            coordinates[i] = new Footprint(xCoords[i],yCoords[i]);
            if (xCoords[i] < west)
                west = xCoords[i];
            if (xCoords[i] > east)
                east = xCoords[i];
            if (yCoords[i] < south)
                south = yCoords[i];
            if (yCoords[i] > north)
                north = yCoords[i];
        }

        // Store minimum enclosing rectangle.
        setBounds(new Footprint(west,south,east-west,north-south));
    }
    // -------------------- Accessor Methods --------------------

    /** Returns an array of footprints representing the vector.
      * @return coordinates of the vector.
      */
```

```
    public Footprint[] getCoords()
    {
        return coordinates;
    }

    /** Reports the type of spatial object.
      * @return Type of vector (POINT, LINE, AREA or VOLUME).
      */
    public int getType()
    {
        return type;
    }
}
```

The constructor requires two arrays of floating point numbers holding the x and y coordinates of the vector object. It also requires a constant informing the class whether the object is a point, line, area or volume. Depending on the type of vector, a series of `if` statements check that there are sufficient coordinates supplied by the arrays to define the object. A simple loop then counts through each coordinate value and creates a new `Footprint` holding each coordinate pair. This loop also keeps track of the outer boundaries of the object as subsequent coordinates are added. Finally, an areal `Footprint` is stored representing the rectangular outer boundary of the object.

Notice that because we have inherited `SpatialObject`, all the relevant constants and the spatial comparison method `compare()` are provided automatically. However, if you remember the code for making a spatial comparison in Chapter 5, it was based on the assumption that the object was rectangular. Thus, this method will only compare the bounding rectangle of the object, not its inner boundary (see Figure 7.2).

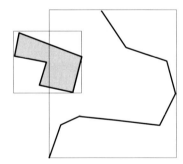

Figure 7.2 Two vector objects and their bounding rectangles. Even though the objects do not intersect, the `compare()` method of each will suggest they do since their bounding rectangles overlap.

We could choose to override the `compare()` method in our `GISVector` class in order to improve the spatial comparison, but as we shall see in Section 8.3, Java already has some classes that can do the hard work for us.

7.2 DYNAMIC GROUPS

Although we can use arrays for creating groups of any primitive or object, we are still presented with a major problem. We have to say how many items are grouped inside our array when we declare it. In some situations, this may be perfectly acceptable (e.g. when creating a new raster based on an existing one), however in others we may not know this information at compile-time.

Consider the `GISVector` class described above. Suppose we wish to create a new vector object to represent part of a road network. We may not know in advance, how many coordinates will be required to represent its entire length. Moreover, we may wish to change the number of coordinates used as new sections of road are added or widened.

We could overcome this problem by declaring our vector array to be of some maximum size that we are unlikely to overflow, but this would waste large amounts of memory. Alternatively, we could create a dynamic group of some kind that can grow and shrink as new items are added or removed.

This section considers some of the classes in Java that allow us to create dynamic groups of data. Such classes are known as *collections* and in Java all use the same *collection framework*. This means that many different types of collection share similar methods for counting through their elements, adding elements, removing elements, etc. It should not surprise you that this is achieved through the use of object-oriented interfaces to ensure consistency between classes.

7.2.1 Java Vectors

Somewhat confusingly for spatial programmers, one of the more widely used dynamic groups in Java is called a `Vector`. This should not be confused with GIS geometrical vectors, although a convenient way to remember what a Java vector does, is that it can store GIS vectors.

A Java vector is equivalent to a one-dimensional array except that

- it can grow and shrink dynamically; and
- elements within the group are addressed using methods rather than [square brackets].

Because `Vector` is not part of the core `java.lang` package, we also have to import the relevant package into our program. Most of the dynamic groups including `Vector` and `TreeMap` (discussed below), are stored in the `java.util` package, so before a class that uses these classes is declared, we need to import the utility package:

```
import java.util.*;
```

To create a Java vector, we declare it like any other class:

```
Vector vectName = new Vector();
```

Because this group can grow and shrink dynamically, we do not need to give it an initial size, although we do have the option if we wish:

```
Vector myVect = new Vector(15);
```

will create a group of 15 (as yet unspecified) objects.

Suppose we wish our Java vector to store a group of `Footprint` objects. How do we fill our group with them? The answer is to use the `add()` method that is part of the `Vector` class:

```
myVect.add(new Footprint(12,6));
```

To add 100 `Footprint` objects to the group, we simply call `add()` 100 times:

```
for (count=0; count<100; count++)
    myVect.add(new Footprint(0,0));
```

If after doing this, we decide that the last `Footprint` object is no longer needed, we can remove it from our group by using the method `removeAt()`:

```
myVect.removeAt(99);
```

Just like an array, each item in the group can be indexed (as in the example above). Unlike an array, the index of any particular element may change dynamically. For example, it is possible to add an item to the middle of the list, shifting everything below it along one place:

```
myVect.add(new Footprint(0,0), 50);
```

Given that the number and position of elements in a Java vector can change, we need to be able to keep track of how big it is. One way of doing this is to use the `size()` method:

```
int numElements = myVect.size();
```

Finally, we need to be able to extract objects from our list, which is done as follows:

```
Footprint location = (Footprint)myVect.get(20);
```

which will set the object `location` to be equal to the 21st point in the vector. Note that we have to *typecast* the result of our vector query since the vector can store any type of object within it.

Collecting Objects Together   161

Vectors and Compatibility with Java 1.1

The methods of the `Vector` class for manipulating its contents are all defined using a common collection *interface* (see 7.2.2 below). Unfortunately, this interface was only included with the Java 2 release. If for any reason, you need to make your code compatible with Java 1.1 (for example to use in an applet), functionality is similar but uses a different set of method names. A comparison of Java 2 and Java 1.1 method names are given below.

Java 2 Vector Method	Java 1.1 Vector Method
`add()`	`addElement()`
`remove()`	`removeElement()`
`insertAt()`	`insertElementAt()`
`get()`	`elementAt()`
`iterator()`	`elements()`

7.2.2 Iterators and the Collections Framework

Elements can be grouped together in different ways depending on their relationship and the way in which we might process the collection. Java provides five basic types of collection interface, illustrated in Figure 7.3. A `Set` consists of an unordered collection of objects with no repetitions, for example, the people surveyed in a census. A `List` consists of an ordered sequence of objects that may contain repetitions. The `Vector` of (x, y) coordinates discussed in the previous section is an example of such a `List`. A `Map` consists of pairs of objects, one acting as an identifier or *key* for the other. The `TreeMap` discussed in the next section is an example of a `Map`. Both `Map`s and `Set`s can also have sorted variations whereby each element can be ordered with respect to other elements in the same collection.

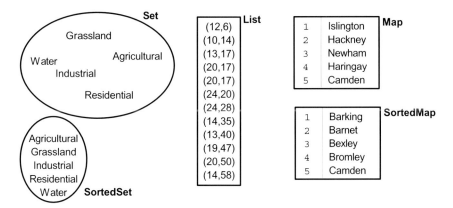

Figure 7.3 Some Java collections.

Despite the differences between the five types of collection provided by Java, they all have features in common. For example, all allow elements to be added, removed and examined. In fact, all these collections implement the same `Collection` interface shown in Table 7.1. This allows the Java programmer to forget about the detailed implementation of each of the collection types and concentrate on the more generic handling of the collection and its elements.

Table 7.1 Methods in the `Collection` interface

Method	Description
`int size()`	Reports number of elements in collection.
`boolean isEmpty()`	Reports whether collection has any elements.
`boolean contains(Object e)`	Reports whether collection contains the given object.
`boolean add(Object e)`	Adds the given object to the collection.
`boolean remove(Object e)`	Removes the given object from the collection
`Iterator iterator()`	Iterates through each element in the collection.
`boolean containsAll(Collection c)`	Reports whether collection contains the given collection of objects.
`boolean addAll(Collection c)`	Adds the given collection of objects to the collection.
`boolean removeAll(Collection c)`	Removes the given collection of objects from the collection.
`boolean retainAll(Collection c)`	Removes all but the given collection of objects from the collection.
`void clear()`	Removes all elements from the collection.
`Object[] toArray()`	Converts the collection into an array.
`Object[] toArray(Object a[]);`	Places the collection in the given array.

By their nature, collections tend to be somewhat dynamic, changing in size and content throughout the running of a program. We saw in the previous section how we can use a loop to count through the elements of a collection one by one. This however can present problems if the size of the collection changes during the operation of the loop. A more flexible and object-oriented solution is to make use of an *iterator*, which processes the elements of a collection one by one, without us having to worry about the detail of how each element is addressed.

All collections (including the `Vector` class we considered above) implement an `iterator()` method that will sequentially retrieve a collection's elements. An `Iterator` has three methods that can be useful when 'traversing' a collection (see Table 7.2).

An iterator works by keeping track of a 'pointer' to a position in the collection. Every time the `next()` method is called, the pointer moves on to the next element in the collection, updating itself as it goes. In this way, we do not need to worry about where or how the next element is retrieved, nor its index value, merely that we can expect that element to be returned by the method when called. The only time `next()` will not retrieve an element is if we have come to the end of the collection. To prevent this from happening we can call the `hasNext()` method to check whether there are any more elements to process.

Table 7.2 Methods in the `Iterator` interface

Method	Description
`Object next()`	Extracts the next element in the collection.
`boolean hasNext()`	Reports whether there are any more elements to extract.
`void remove()`	Removes the currently selected element from the collection.

Bringing this all together, it is a relatively simple task to use a 'while-do' loop to process all the elements in a collection. For example, the following will remove all elements in a collection of `Strings` (whether a `List`, `Set` or `SortedSet`) that are equal to 'not known'.

```
Collection names;    // Assume names is some collection of Strings.

Iterator I = names.iterator();
while (i.hasNext())
{
    String element = (String)i.next();
    if (element.equals("not known"))
        i.remove();
}
```

7.2.3 An Improved GIS Vector Model using Dynamic Collections

We can make our `GISVector` class much more flexible by incorporating dynamic collections in the storage of geometry. If we consider the storage of coordinates in `GISVector`, it should be apparent that some kind of `List` is the most suitable form of collection since coordinates are ordered and may be repeated in a collection. The type of `List` we shall use is the (Java) `Vector` class as this contains all the necessary behaviour required to manipulate a list of coordinates. In fact we only have to make a few minor modifications to our original `GISVector` class to do this. We can leave the array-based constructor developed in Section 7.1 and merely change the internal representation of coordinates from an array to a `Vector`. This then gives us the flexibility to create an 'empty constructor' and a method for adding coordinates to the GIS vector. The changes are shown below.

```
package jwo.jpss.spatial;  // Part of the spatial modelling package.
import java.util.*;        // For Java Vector class.

//    ****************************************************************
/** Models a GIS vector object.
  * @author Jo Wood
  * @version 2.3, 7th October, 2001
```

```java
     */
//   ****************************************************************
     public class GISVector extends SpatialObject
     {
         // -------------------- Object variables --------------------

         private Vector coordinates;    // Vector geometry array.
         private int type;              // Type of vector object.

         // --------------------- Constructor ---------------------

         /** Creates an empty GIS vector object.
           */
         public GISVector()
         {
            type = POINT;
            coordinates = new Vector();
            setBounds(new Footprint(0,0,0,0));
         }

         /** Creates GIS vector object with the given x and y coords.
           * @param xCoords Array holding x-coordinates of object.
           * @param yCoords Array holding y-coordinates of object.
           * @param type Type of vector (POINT, LINE, AREA, VOLUME).
           */
         public GISVector(float[] xCoords, float yCoords[], int type)
         {
            // Initialise vector information and check integrity.

            int numCoords = xCoords.length;
            this.type = type;
            coordinates = new Vector();

            // Integrity code as previous version.
            // ...

            // Store the coordinates and calculate bounds of object.
            for (int i=0; i<numCoords; i++)
            {
                coordinates.add(new Footprint(xCoords[i],yCoords[i]));

                // Bounding rectangle code as previous version
                // ...
            }
         }

         // ------------------------ Methods ------------------------

         /** Adds the given footprint to the vector's coordinates and
           * updates the bounding area.
           * @param footprint Coordinates to add.
           */
         public void addCoords(Footprint footprint)
         {
            coordinates.add(footprint);
            setBounds(getUnionMER(footprint));
         }
```

Collecting Objects Together

```
// -------------------- Accessor Methods --------------------

/** Returns array of footprints representing the vector object.
  * @return coordinates of the vector.
  */
public Footprint[] getCoords()
{
    return (Footprint[])coordinates.toArray();
}

// -------------------- Mutator Methods --------------------

/** Sets the type of vector.
  * @param Type of vector (POINT, LINE, AREA or VOLUME).
  */
public void setType(int type)
{
    this.type = type;
}
}
```

Since this class now allows coordinates to be added dynamically, so the bounding rectangle of the object may change accordingly. The `addCoords()` method calls a new method `getUnionMER()` which will create a new bounding rectangle based on the union of the existing bounds and the new footprint being added. Since this operation might be useful to a range of possible spatial objects, it makes sense to place the code in the `SpatialObject` class making it available to any of its 'children'.

```
/** Calculates bounds of the union of given footprint and this MER.
  * @param otherFp Footprint of object to union with this MER.
  */
public Footprint getUnionMER(Footprint otherFp)
{
    // Store Minimum enclosing rectangles of both footprints.
    float minX,minY,maxX,maxY;
    minX = Math.min(footprint.getXOrigin(),otherFp.getXOrigin());
    minY = Math.min(footprint.getYOrigin(),otherFp.getYOrigin());
    maxX = Math.max(footprint.getXOrigin()+footprint.getMERWidth(),
                    otherFp.getXOrigin()+otherFp.getMERWidth());
    maxY = Math.max(footprint.getYOrigin()+footprint.getMERHeight(),
                    otherFp.getYOrigin()+otherFp.getMERHeight());

    return new Footprint(minX,minY,maxX-minX,maxY-minY);
}
```

A `GISVector` can be used for storing the geometry of a feature such as a river, building outline, or city location. However most useful models of spatial information will consist of many of these features assembled together. It will therefore be useful to be able to group these `GISVector` objects together into a `VectorMap` object capable of representing more complex models of space (see Figure 7.4). Once again it makes sense to use a dynamic collection to assemble these objects.

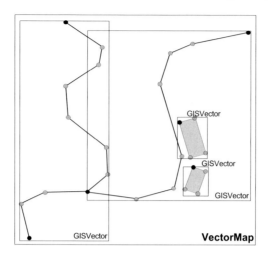

Figure 7.4 A collection of `GISVector`s assembled in a `VectorMap`.

```
package jwo.jpss.spatial;  // Part of the spatial modelling package.
import java.util.*;        // For the Java collection framework.

// ***************************************************************
/** A collection of GIS vectors that model object boundaries.
  * @author Jo Wood
  * @version 2.3, 6th October, 2001
  */
// ***************************************************************

public class VectorMap extends SpatialObject
{
    // -------------------- Object variables --------------------

    private Vector gisVectors;    // Collection of GIS vectors.

    // -------------------- Constructors ----------------------

    /** Creates an empty vector map with default characteristics.
      */
    public VectorMap()
    {
        gisVectors = new Vector();
        setBounds(new Footprint(0,0,1,1));
    }

    /** Creates a vector map from the given GIS vector.
      * @param gisVector GIS vector from which to create map.
      */
    public VectorMap(GISVector gisVector)
    {
        // Add vector object to map.
        gisVectors = new Vector();
        add(gisVector);
```

```java
        // Store its bounding rectangle.
        updateBounds();
    }
    // ---------------------- Methods --------------------------

    /** Adds the given GIS vector object to the map.
      * @param gisVector GIS vector object to add.
      */
    public void add(GISVector gisVector)
    {
        gisVectors.add(gisVector);
        updateBounds();
    }

    /** Removes the given GIS vector object from the map. If the
      * map does not contain the object, it is unchanged.
      * @param gisVector GIS Vector object to remove.
      * @return True if vector object removed.
      */
    public boolean remove(GISVector gisVector)
    {
        return gisVectors.remove(gisVector);
    }

    /** Returns a collection of GIS vectors stored in this map.
      * @return Collection of GIS vectors.
      */
    public Vector getGISVectors()
    {
        return gisVectors;
    }

    /** Reports the details of this vector map.
      * @return Summary of this vector map.
      */
    public String toString()
    {
       return "Vector map with "+gisVectors.size()+" objects and "+
              getBounds().toString();
    }

    // ------------------ Implemented Methods --------------------

    /** Reports the type of spatial model.
      * @return Type of spatial model (VECTOR_2D, VECTOR_3D etc).
      */
    public int getType()
    {
        return VECTOR_2D;
    }

    // -------------------- Private Methods --------------------

    /** Updates the minimum enclosing rectangle as that enclosing
      * all GIS Vectors within this map.
      */
    private void updateBounds()
    {
```

```
        // Check there are some GIS vectors to examine.
        if (gisVectors.size() == 0)
            return;

        // Initialise map bounds with those of first GIS vector.
        Iterator i = gisVectors.iterator();
        GISVector gisVector = (GISVector)i.next();
        setBounds(gisVector.getBounds());

        // Search through remaining objects updating bounds.
        while (i.hasNext())
        {
            gisVector = (GISVector)i.next();
            setBounds(getUnionMER(gisVector.getBounds()));
        }
    }
}
```

This class makes use of more the `iterator()` method of a Java collection using it to check through each of the GIS vector objects in turn in order to find the bounding rectangle of the whole map. It also uses the `add()` and `remove()` methods of the `Collection` interface to allow GIS vector objects to be added to and removed from the map.

7.2.4 The TreeMap and HashMap

`TreeMap`s and `HashMap`s are implementations of the `Map` interface allowing each element in a collection to be uniquely identified by a *key* of some type. This key can be any Java object, but is often a `String` object that contains some uniquely identifiable text. `TreeMap`s ensure that keys are sorted in some identifiable way, making the process of searching though a list of keys potentially more efficient.

One way of thinking of both of these collection types is to consider them as representations of tables much as you might when considering a relational database. In both cases, one of the columns in the table acts as a key allowing the unique identification of any row within the table.

Database Connectivity

Although it would be possible to create a simple relational database using HashMaps, any serious use of Relational Database Management Systems or the Structured Query Language (SQL) would be more likely to use the *Java Database Connectivity* (JDBC) library. This allows Java programs to connect with a range of commercially produced database management systems. See `java.sun.com/j2se/1.4/docs/guide/jdbc` for further details.

Consider how we might use a `Map` to create a simple *gazetteer* – a resource that allows us to relate a named feature (for example, a city, a building, or a mountain) to a location (see Figure 7.5).

Collecting Objects Together

Key		Value
City Hall	civic	257549,9857990
Central Railway Station	transport	258300,9857434
Chogoria Road	transport	258728,9856197
Nairobi Hospital	civic	255650,9856930
nbi-217	mapsheet	257100,9857000,1900,1100

```
      String                    Footprint
```

Figure 7.5 A simple Gazetteer table using two classes to store the keys and values.

To create a `HashMap` to represent the gazetteer table, we first (as always) create an object from the class:

```
HashMap gazetteer = new HashMap();
```

We can add and remove entries to the table using the methods `put()` and `remove()` methods. To add an object, we specify both the key to use as an identifier and the object to be stored. So for example,

```
gazetteer.put("City Hall, Civic"),
           new Footprint(257549,9857990));
```

will add a new `String` object as an identifying key and a new `Footprint` as its corresponding value.

To retrieve an object, we must both supply the key to identify the relevant table row as well as typecast the object to be retrieved (as we did for the `Vector` class). So, for example,

```
Footprint location =
           (Footprint)gazetteer.get("City Hall, Civic");
```

The `HashMap` implements the unsorted version of the `Map` interface, meaning that the order in which the keys are stored in the table is undetermined and should not be relied upon for searching. If we were to create a gazetteer with many thousands of entries, this would become an increasingly inefficient way in which to handle the text keys. We can improve the gazetteer model by using the `TreeMap` class instead, which relies upon the keys being sorted in some way.

To use a `TreeMap`, the class used as the key must implement a `Comparable` interface – one that contains a method `compareTo()` that will report the relative order of the object and another of the same type. The `String` class used in the example above already implements this method and would allow single text entries to be placed in alphabetic order.

In the case of the gazetteer, the key actually consists of two parts – the name of the feature, and its classification. We can therefore improve our ability to search through the gazetteer by creating a `Comparable TextEntry` class that sorts first by feature name, then by feature type.

```java
/** Stores a gazetteer text entry. Consists of two text fields
  * describing the spatial object and its type.
  */
public class TextEntry implements Comparable
{
    // -------------------- Object variables --------------------

    private String name,type;    // Name of entry and its type.

    // --------------------- Constructor ---------------------

    /** Creates a text entry with a given name and type.
      * @param name Name of object in gazetteer.
      * @param type Type of object in gazetteer.
      */
    public TextEntry(String name, String type)
    {
        this.name = name;
        this.type = type;
    }

    // -------------------- Accessor Methods --------------------

    /** Reports the name of the gazetteer object.
      * @return Name of gazetteer object.
      */
    public String getName()
    {
        return name;
    }

    /** Reports the type of the gazetteer object.
      * @return type of gazetteer object.
      */
    public String getType()
    {
        return type;
    }

    /** Reports the contents of the text entry in a form suitable
      * for display.
      * @return Contents of the text entry object.
      */
    public String toString()
    {
        return getName()+","+getType();
    }

    // ----------------- Implemented Methods --------------------
    /** Compares this object with another. Used for placing
      * text entries in alphabetic order.
      * @param o Object with which to compare this one.
      * @return Comparative alphabetic order of the two objects
      */
    public int compareTo(Object o)
```

```
        {
            TextEntry otherObj = (TextEntry)o;
            int placeCmp = name.compareTo(otherObj.getName());

            if (placeCmp == 0)
                return type.compareTo(otherObj.getType());
            else
                return placeCmp;
        }
}
```

The compareTo() method returns a number indicating the relative sorting order of the object when compared with another of the same type. If the text of the two objects is identical then a value of 0 is returned. If the object is alphabetically earlier than the given object, negative value is returned, otherwise a positive number is reported.

Using TextEntry and Footprint to represent the key and value in our TreeMap, it now becomes a relatively simple task to add a new Gazetteer class to the jwo.jpss.spatial package.

```
package jwo.jpss.spatial;  // Part of the spatial modelling package.
import java.util.*;        // For the Java collections framework.

//   *****************************************************************
/** Stores a Gazetteer table and allows simple searching.
  * @author    Jo Wood.
  * @version   1.3, 7th October, 2001.
  */
//   *****************************************************************

public class Gazetteer
{
    // -------------------- Object variables ---------------------

    private TreeMap entries;        // Gazetteer entries.

    // --------------------- Constructors -----------------------

    /** Initialises the gazetteer.
      */
    public Gazetteer()
    {
        entries = new TreeMap();
    }

    // ---------------------- Methods -------------------------

    /** Adds an entry to the gazetteer. Each entry should consist
      * of a location, a name and a feature type.
      * @param fp Location/extent of the feature.
      * @param name Name of the feature.
      * @param type Feature classification.
      */
    public void addEntry(Footprint fp, String name, String type)
    {
```

```java
            TextEntry key = new TextEntry(name,type);
            entries.put(key,fp);
    }

    /** Reports location corresponding to the given feature name
      * @param Name of feature to search for.
      * @return Location of given feature or null if none found.
      */
    public Footprint getLocation(String featureName)
    {
        Iterator keys = entries.keySet().iterator();

        while (keys.hasNext())
        {
            TextEntry place = (TextEntry)keys.next();

            if (place.getName().equals(featureName))
                return (Footprint)entries.get(place);
        }

        // No match found if we get this far.
        return null;
    }

    /** Reports all current gazetteer entries as a 2D array.
      * Useful for importing into a JTable.
      * @return Gazetteer entries as 2D array of n rows by 2 cols.
      */
    public String[][] getEntryArray()
    {
        String [][] gazArray = new String[entries.size()][2];
        Iterator keys = entries.keySet().iterator();
        int row=0;

        while (keys.hasNext())
        {
            TextEntry place = (TextEntry)keys.next();
            gazArray[row][0] = place.getName();
            gazArray[row][1] = place.getType();
            row++;
        }

        return gazArray;
    }

    /** Reports the gazetteer entries that contain the given string
      * as a 2D array. Useful for importing into a JTable.
      * @param text Text to search for.
      * @return All gazetteer entries that match the given string.
      */
    public String[][] getEntryArray(String text)
    {
        Vector matchedEntries = new Vector();

        String [][] gazArray;
        String srchStr = text.toLowerCase();
        Iterator keys = entries.keySet().iterator();

        // Search for string and store matched entries.
```

```
        while (keys.hasNext())
        {
            TextEntry place = (TextEntry)keys.next();

            if (place.getName().toLowerCase().indexOf(srchStr) >=0)
                matchedEntries.add(place);
        }

        // Convert matched entries into an array.
        int numMatches = matchedEntries.size();
        gazArray = new String[numMatches][2];
        Iterator matchedKeys = matchedEntries.iterator();
        int row=0;

        while (matchedKeys.hasNext())
        {
            TextEntry place = (TextEntry)matchedKeys.next();
            gazArray[row][0] = place.getName();
            gazArray[row][1] = place.getType();
        }

        return gazArray;
    }
    /** Reports the number of entries in the gazetteer.
      * @return Number of entries in the gazetteer.
      */
    public int getNumEntries()
    {
        return entries.size();
    }
}
```

7.3 CASE STUDY: FEEDING ANT COLONIES

If you recall from the previous chapter, our ant simulation model is capable of displaying a window containing a garden, nest and a single roaming ant. We can use the idea of the dynamic collection to expand this model to accommodate many ants each of which can roam independently around the garden. In order to keep this new ant colony alive, we shall add life-sustaining food to the garden and allow our ants to breed.

7.3.1 Adding Food

We have already given our ants the ability to metabolise their food supply as well as gain energy by eating food. What we have not done is to place any food in the garden for the ants to eat. We should be able to place food in the garden just like any other object (ants and nest). Like the other objects in the garden, we will also need to get food to draw itself at an appropriate location. So do we create a new food class from scratch or should we inherit or implement another class or interface? A logical starting point will be to create a new class `FoodSource` that inherits `SpatialObject`, just as `Garden`, `Nest` and `Animal` did.

Given that `FoodSource` inherits the functionality of `SpatialObject`, all we need to add is some drawing code and methods to allow the amount of food stored in the food source to be queried and changed.

```
package jwo.jpss.ants;     // Part of the ant simulation package.
import java.awt.*;         // For drawing the food.
import jwo.jpss.spatial.*; // For spatial classes.

// ***********************************************************
/** Class for defining a food source.
  * @author  Jo Wood
  * @version 1.4, 10th October, 2001.
  */
// ***********************************************************
public class FoodSource extends SpatialObject implements Drawable
{
    // ------------- Class and object variables -------------

    private int foodLevel;            // Amount of available food.
    private int x,y,width,height;     // Drawable bounds.

    // ------------------ Constructors --------------------

    /** Creates a food source with a given location and value.
      * @param x x location of the food source.
      * @param y y location of the food source.
      * @param fLevel Amount of available food.
      */
    public FoodSource(float x, float y, int fLevel)
    {
        super(new Footprint(x,y,1 + fLevel/1000,1+fLevel/1000));
        setBounds(getBounds());
        this.foodLevel = foodLevel;
    }

    // ---------------------- Methods ----------------------

    /** Returns the level of food left in the source.
      * @return The level of food left in the source.
      */
    public int getFoodLevel()
    {
        return foodLevel;
    }

    /** Allows a given amount of food to be removed. If there is
      * not enough food left, remaining food will be removed.
      * @return foodToRemove Amount of food to remove.
      * @return Amount of food actually removed from the source.
      */
    public int removeFood(int foodToRemove)
    {
        int foodRemoved;

        // Reduce food supply.
        if (foodLevel >= foodToRemove)
            foodRemoved = foodToRemove;
        else
```

Collecting Objects Together

```
                foodRemoved = foodLevel;

        // Update size of food source.
        foodLevel -= foodRemoved;
        Footprint newSize = getBounds();
        newSize.setMERWidth(1 + foodLevel/1000);
        newSize.setMERHeight(1 + foodLevel/1000);
        setBounds(newSize);

        return foodRemoved;
    }

    /** Draws the food source.
      * @param g Graphics context in which to draw the food source.
      */
    public void paint(Graphics g)
    {
        g.setColor(new Color(0,155,10));
        g.fillRect(x,y,width,height);

        g.setColor(Color.black);
        g.drawRect(x,y,width,height);
    }

    /** Sets the bounding rectangle of this object.
      * @param bounds New spatial footprint of this food source.
      */
    public void setBounds(Footprint bounds)
    {
        super.setBounds(bounds);

        // Store drawable bounds.
        x = Math.round(bounds.getXOrigin());
        y = Math.round(bounds.getYOrigin());
        width = Math.round(bounds.getMERWidth());
        height = Math.round(bounds.getMERHeight());
    }
}
```

This class does little more than keep track of the number of food units within a food source, allowing it to be reduced with the method `removeFood()`. It also draws itself in proportion to the number of food units it contains (1 pixel = 1000 food units). We shall see how we can add these `FoodSources` to the garden in the next section.

7.3.2 Adding Collections of Objects

Our previous simulation design consisted of a `Garden` containing exactly one `Nest` which itself produced 1 `Ant`. We can improve our model substantially by creating dynamic collections of `Nests` and `Ants` along with a collection of `FoodSources`. Each of these can conveniently be stored in a Java `Vector` object.

The obvious place to store the collection of food sources and nests is in the

Garden class. The constructor simply adds two new Nest objects to a Vector of nests, and 40 FoodSources to a Vector of foodSources.

Constructor from Garden

```
private Vector nests, newNests;      // A collection of ant nests.
private Vector foodSources;          // A collection of food sources.

// ---------------------- Constructor -------------------------

/** Creates a garden with ants' nests and food.
  * @param footprint Spatial footprint of garden.
  */
public Garden(Footprint footprint)
{
    super(footprint);

    // Store garden footprint.
    Footprint fp = getBounds();
    bounds = new Rectangle((int)fp.getXOrigin(),
                           (int)fp.getYOrigin(),
                           (int)fp.getMERWidth(),
                           (int)fp.getMERHeight());

    // Add 2 ants' nests to the garden.
    nests = new Vector(2);
    float nestY = (fp.getMERHeight()/2)-10;
    float nestX = (fp.getMERWidth()/3);
    nests.add(new Nest(nestX-10,nestY));
    nests.add(new Nest(2*nestX-10,nestY));

    // Add food sources to garden.
    foodSources = new Vector(40);
    for (int i=0; i<40; i++)
    {
        int x = (int)((bounds.width-30)*Math.random());
        int y = (int)((bounds.height-30)*Math.random());
        foodSources.add(new FoodSource(x,y,10000));
    }
}
```

We can then use an *iterator* to cycle through these collections in order both to display them within the garden and to get each one to evolve.

paint() *method from the* Garden *class*

```
/** Draws the garden and all contained within it using the
  * given graphics context.
  * @param g Graphics context to draw to.
  */
public void paint(Graphics g)
{
    // Draw garden.
    g.setColor(new Color(200,220,200));
    g.fillRect(bounds.x, bounds.y, bounds.width, bounds.height);
```

```
    // Draw nests.
    Iterator i = nests.iterator();
    while (i.hasNext())
    {
        Nest nest = (Nest)i.next();
        nest.paint(g);
    }

    // Draw food sources.
    i = foodSources.iterator();
    while (i.hasNext())
    {
        FoodSource foodSource = (FoodSource)i.next();
        foodSource.paint(g);
    }
}
```

evolve() method from the Garden class

```
/** Let the garden grow for one time unit.
  */
public void evolve()
{
    // Remove any food sources that have run out of food.
    Iterator i = foodSources.iterator();
    while (i.hasNext())
    {
        FoodSource food = (FoodSource)i.next();
        if (food.getFoodLevel() == 0)
            i.remove();
    }

    // Possibly add a food source at random location.
    if (Math.random() < 0.02)
    {
        int x = (int)((bounds.width-30)*Math.random());
        int y = (int)((bounds.height-30)*Math.random());

        foodSources.add(new FoodSource(x,y,10000));
        gListener.redrawGraphics();
    }

    // Let the ants' nest evolve.
    i = nests.iterator();
    while (i.hasNext())
    {
        Nest nest = (Nest)i.next();
        nest.evolve();
    }

    // Check to see if any new nests have been requested.
    if (newNests.size() > 0)
    {
        nests.addAll(newNests);
        newNests.clear();
    }
}
```

This method makes substantial use of the `Collection` interface using an `Iterator` to cycle through the collections of nests and food sources. If any of the food sources have had their entire food supply depleted, they are removed from the collection using `Iterator`'s `remove()` method. The method also makes use of the `Collection`'s bulk operator `addAll()`, which adds a new collection of nests to the garden if any of the ants have been sufficiently productive (see 7.3.5 *Adding a Queen* below).

The remaining collection to add to our simulation is the `Vector` of `Ants` associated with each `Nest` object. This is coded in much the same way as the `Nests` and `FoodSources` in `Garden`, using an `Iterator` to cycle through each ant in turn both to draw and to evolve it.

paint() method from the Nest class

```java
/** Draws the nest.
 * @param g Graphics context in which to draw the nest.
 */
public void paint(Graphics g)
{
    // Give nest a colour if it is alive.
    if (isAlive())
    {
        g.setColor(new Color(155,155,100));
        g.fillOval(bounds.x, bounds.y, bounds.width, bounds.height);
    }
    g.setColor(Color.black);
    g.drawOval(bounds.x, bounds.y, bounds.width, bounds.height);

    // Draw ants associated with this nest.
    Iterator i = ants.iterator();
    while (i.hasNext())
    {
        Ant ant = (Ant)i.next();
        ant.paint(g);
    }
}
```

7.3.3 Giving Ants Sentience

We now have a garden stocked with food, a couple of nests and roaming ants, but so far the ants lack any awareness of their surroundings other than the boundary of the garden. If we wish ants to feed when they stumble across a food source, we must give them a mechanism for recognising what is in the garden. For the moment, we will allow them to identify food sources and the nest in which they were born.

How do we get an ant to recognise a food source or a nest? At first glance this seems like a difficult problem as our object-oriented design has encouraged us to create classes that are as independent of one another as possible. There is nothing

Collecting Objects Together

stored by the Ant class that makes direct reference to any FoodSource objects. What we need is access to a class that knows about the existence of all the food (and nests). We created just such a class when we added food to the Garden. However, we still have a problem because while the Garden contains reference to Ants, the Ant class does not contain any reference to Garden. It does however contain reference to the GraphicsListener we used to link all our drawable classes together. This provides the key to communication between objects in the garden.

We need to add a new method to our GraphicsListener interface that allows us to identify what objects are present in a given location. Any class that has access to a GraphicsListener can then find out what is contained in its immediate surroundings. The abstract method is simply defined with respect to an arbitrary SpatialObject.

Abstract method in the GraphicsListener interface

```
/** Reports the list of SpatialObjects associated with the given
  * spatial object (within, matching, overlapping or containing).
  * @param spatialObject Spatial object with which to compare.
  * @return List of spatial objects in contact with the given one.
  */
public Vector objectsAt(SpatialObject spatialObject);
```

When implemented, this method should return a list of objects (stored in a Vector) that fall within, overlap, match or contain the footprint of the given spatial object. Currently, the only class that is obliged to implement this method is the one that implements the GraphicsListener interface, namely GardenPanel contained within AntApplication. We also know that the only class capable of actually finding out what is at a given location is the Garden class. So, GardenPanel simply delegates the query to Garden to do the hard work.

objectsAt() method inside GardenPanel

```
/** Reports the list of SpatialObjects assoicated with the given
  * spatial object (within, matching, overlapping or containing).
  * @param spObject spatial object with which to compare.
  * @return List of spatial objects in contact with the given one.
  */
public Vector objectsAt(SpatialObject spObject)
{
    return garden.objectsAt(spObject);
}
```

objectsAt() method inside Garden

```
/** Reports the list of SpatialObjects associated with the given
  * spatial object (within, matching, overlapping or containing).
  * @param spObject Spatial object with which to compare.
  * @return List of spatial objects in contact with the given one.
```

```java
    */
public Vector objectsAt(SpatialObject spObject)
{
    Vector spatialObjects = new Vector();

    // If garden is separate from footprint, there can be no other
    // objects at this footprint.
    if (compare(spObject) == SEPARATE)
        return spatialObjects;

    // Add any nest items that are connected to footprint.
    Iterator i = nests.iterator();
    while (i.hasNext())
    {
        Nest nest = (Nest)i.next();
        if (nest.compare(spObject) != SEPARATE)
            spatialObjects.add(nest);
    }

    // Add any food sources that are connected to footprint.
    i = foodSources.iterator();
    while (i.hasNext())
    {
        FoodSource foodSource = (FoodSource)i.next();

        if (foodSource.compare(spObject) != SEPARATE)
            spatialObjects.add(foodSource);
    }

    // Add this garden to the end of the list.
    spatialObjects.add(this);

    return spatialObjects;
}
```

This method simply iterates though the lists of nests and food sources, building up a list of any that are spatially connected with the supplied spatial object. The spatial comparison is handled by the `compare()` method already coded as part of the `SpatialObject` class.

To give an ant sentience, it simply calls the `ObjectsAt()` method of its `GraphicsListener` giving itself as a parameter, and processes the resulting list of objects.

Ant sentience code

```java
/** Let the ant go about its business for one time unit.
  */
public void evolve()
{
    metabolise(METABOLIC_RATE);

    if (isAlive())
    {
        // Find out if the ant is sitting on anything.
        Iterator i = gListener.objectsAt(this).iterator();
```

```
            while (i.hasNext())
            {
                SpatialObject spObject = (SpatialObject) i.next();

                // If ant is sitting on some food, eat it.
                if (spObject instanceof FoodSource)
                {
                    // Attempt to eat food at feeding rate.
                    FoodSource foodSource = (FoodSource)spObject;
                    int foodRemoved=0;

                    if (getFoodLevel()+FEED_RATE <= maxFood)
                        foodRemoved = foodSource.removeFood(FEED_RATE);

                    if (foodRemoved > 0)
                    {
                        eat(foodRemoved);
                        return;
                    }
                }

                // If ant finds itself at its own nest, the nest is
                // alive and ant has enough food, give half to nest.
                if (spObject == nest)
                {
                    int foodLevel = getFoodLevel();
                    goingHome = false;

                    if ((foodLevel>DONATION_LEVEL) && (nest.isAlive()))
                    {
                        nest.addFood(this,foodLevel/2);
                        metabolise(foodLevel/2);
                        return;
                    }
                }
            }
            // Remaining ant movement code here.
            // ...
        }
    }
```

If the ant is alive, it searches through the list of objects with which it shares a spatial location. All of the objects returned by objectsAt() will be SpatialObjects, but they may be of different types. The inquisitive ant checks whether it has come across *any* FoodSource using a new Java operator – instanceof. As the name suggests, this operator tests to see whether the given SpatialObject is an instance of FoodSource. If the ant is not full up, it will eat the food at its own feeding rate (defined by the static constant FEED_RATE).

Any excess food carried by the ant will be stored for later metabolism or for supplying its nest. We can get an ant to indicate how much excess food it is carrying by modifying its paint() method to draw its stored food. In this case, we will simply draw a coloured square in proportion to the amount of food carried (indicated by the variable foodSize below).

```
/** Draws the ant using the given graphics context.
  * @param g Graphics context to draw to.
  */
public void paint(Graphics g)
{
    Footprint fp = getBounds();
    int x = Math.round(fp.getXOrigin());
    int y = Math.round(fp.getYOrigin());
    int width = Math.round(fp.getMERWidth());
    int height = Math.round(fp.getMERHeight());

    g.setColor(colour);
    g.fillOval(x,y,width,height);

    if (getFoodLevel() > 1000)
    {
        g.setColor(new Color(0,155,100));
        g.fillRect(x+width,y, foodSize+1, foodSize+1);
        g.setColor(colour);
        g.fillRect(x+width,y, foodSize+1, foodSize+1);
    }
}
```

7.3.4 Ant Evolution

Our next task in this case study is to give the ants the ability to breed and evolve different strategies for survival. In the first instance, we will limit the strategies to a few simple characteristics encoded in the ant's 'genes'. These are shown in Table 7.3. New ants will inherit these genes as they are created.

Table 7.3 Ant genes

Gene	Characteristic
homingInstinct	Probability that ant will return to its nest at any given time interval.
straightSteps	Number of straight steps taken before changing direction.
maxFood	Maximum amount of food that can be carried by the ant.
colour	Colour of ant.

Breeding ants allows us to introduce a relatively new concept in computational problem solving – that of evolutionary programming. The ants are placed in an environment where some die while others survive and go on to produce offspring. The offspring inherit the 'genetic' properties of their parents with perhaps some minor mutation. The result after many generations is the evolution of ants with behaviour more suited to the prevailing environment, since any ants with unsuitable behaviour are less likely to survive to produce offspring. This approach is easily modelled using the object-oriented paradigm built around the ideas of inheritance, state and behaviour.

Movement is similar to the simple random model adopted in Section 6.4. where the ant moves in a randomly determined straight line for straightSteps cycles. We will add two new variations to the model here. First, that the speed of movement is influenced by the amount of food carried by the ant (greater food levels slow the ant down). Second, the ant has a small probability that it will return to its nest, determined by the `homingInstinct` gene.

Ant movement code

```java
/** Let the ant go about its business for one time unit.
  */
public void evolve()
{
    metabolise(METABOLIC_RATE);

    if (isAlive())
    {
        // Feeding behaviour code here.
        // ...

        // Change direction if any has walked sufficient steps.
        if (numSteps%straightSteps == 0)
           changeAntDirection();

        move(xDir,yDir);

        // Only move if ant is within bounds.
        if (gListener.canDraw(this))
            numSteps++;
        else
        {
            // Change direction if cannot move.
            move(-xDir,-yDir);
            changeAntDirection();
        }
    }
}
/** Sets a new direction for ant to walk in.
  */
private void changeAntDirection()
{
    // No need to change direction if ant is already going home.
    if (goingHome)
        return;

    // Speed of ant depends on how much it is carrying.
    float speed = 10f/(5f+getFoodLevel()/1000f);

    // If ant decides to go home, calculate new direction.
    if ((Math.random() < homingInstinct) ||
        (maxFood - getFoodLevel() < FEED_RATE))
    {
        goingHome = true;
        Footprint nestFP = nest.getBounds(),
                  antFP  = getBounds();
        float dx =(nestFP.getXOrigin()+ nestFP.getMERWidth()/2) -
                  (antFP.getXOrigin() + antFP.getMERWidth()/2);
        float dy =(nestFP.getYOrigin()+ nestFP.getMERHeight()/2) -
                  (antFP.getYOrigin() + antFP.getMERHeight()/2);
```

```
            float scaling = (float)Math.sqrt(dx*dx + dy*dy)/speed;

            if (scaling != 0)
            {
                xDir = dx/scaling;
                yDir = dy/scaling;
            }
        }
        else    // Otherwise calculate new random direction.
        {
            xDir = (float)(Math.random()*2-1)*speed;
            yDir = (float)(Math.random()*2-1)*speed;
        }
    }
}
```

In what sense are the ants' behavioural characteristics 'genes' rather than normal variables representing their state? The difference becomes clearer when we allow our ants to breed. If an ant ever produces offspring, its offspring will inherit the four genes of its parent with a minor mutation. The form of the mutation can be represented as

m = X(r – 0.5)

where **m** is the amount of mutation, **X** is the maximum mutation and **r** is a random function between 0 and 1. The result is a mutation that can up to + or –**X/2**. By default the value of **X** for a given gene is set to 5% of its initial maximum value.

To implement inheritance with mutation, we simply add a new constructor to our Ant class that takes an Ant object as one of its arguments, extracts the genes of that object and mutates them according to the rules above.

```
// Declare immutable ant 'laws' as class variables.
                        // Proportion of mutation.
private static final float MUTATION = 0.05f;
                        // Maximum straight steps.
private static final int   MAX_STRAIGHT_STEPS = 20;
                        // Initial maximum homing probability.
private static final float MAX_HOMING_INSTINCT = 0.1f;
                        // Initial maximum food capacity.
private static final int   MAX_FOOD = 10000;
                        // Width of ant.
private static final int   WIDTH = 6;
                        // Height of ant.
private static final int   HEIGHT = 3;
                        // Food per cycle that can be eaten.
private static final int   FEED_RATE = 100;
                        // Food level at which ant gives to nest.
private static final int   DONATION_LEVEL = 1000;
                        // Rate at which ant consumes food.
private static final int   METABOLIC_RATE = 1;


// -------------------- Constructors --------------------

/** Creates an ant with similar characteristics of the given
```

```
 * parent. Applies small genetic mutation to inherited genes.
 * @param foodLevel Initial food level of the ant.
 * @param parent Ant supplying inheritable characteristics.
 * @param x Initial x location of the ant.
 * @param y Initial y location of the ant.
 * @param nest Nest in which this ant was born.
 */
public Ant(int foodLevel, Ant parent, float x, float y, Nest nest)
{
    super(foodLevel, new Footprint(x,y,WIDTH,HEIGHT));
    this.nest = nest;

    float mutation;
    numSteps = (int)(Math.random()*10);
    foodSize = 0;

    // Inherit straight steps with mutation.
    mutation = MAX_STRAIGHT_STEPS*MUTATION*(Math.random()-0.5f);
    straightSteps = Math.round(parent.getStraightSteps()+mutation);

    if (straightSteps <1)
        straightSteps = 1;

    // Inherit homing instinct with mutation.
    mutation = MAX_HOMING_INSTINCT*MUTATION*(Math.random()- 0.5f);
    homingInstinct = parent.getHomingInstinct()+mutation;

    // Inherit maximum food capacity.
    mutation = (float)(MAX_FOOD*MUTATION*(Math.random() - 0.5f));
    maxFood = Math.round(parent.getMaxFood()+mutation);

    // Inherit parent colour with mutation.
    colour = parent.getColour();
    if (Math.random() <MUTATION)
    {
        if (Math.random() < 0.5)
            colour = colour.brighter();
        else
            colour = colour.darker();
    }
}
```

The constructor finds out each of the genetic characteristics of the 'parent ant' and adds or subtracts a small value to it. Where the genes have some fixed maximum or minimum (such as the minimum number of straight steps), a limit is placed on possible mutated values. Those characteristics that cannot change such as the size of the ant and its metabolic rate, have been stored as static constants. This helps to distinguish those characteristics that are immutable from those that might vary.

So, now our ants wander around the garden, periodically changing direction, feeding when they come across any food and occasionally returning to their nest with the food they have collected. We have also given each ant the ability to pass its genes on to its progeny, but have not yet allowed the ants to breed. To produce new ants, we can imagine a hardworking queen ant in each nest that will produce baby ants whenever it has enough food. Where does it get its food from? From the roaming worker ants returning the nest with a supply of freshly gathered produce.

We can model this behaviour quite simply if we allow `Nest` to inherit the `Animal` class rather than the `SpatialObject` class directly. By doing so, `Nest` will still retain all the state and behaviour of a spatial object (`Animal` inherits `SpatialObject`), but will also have its own food level, metabolic rate etc. We can then modify the Nest's `evolve()` method to produce new ants only when the nest has accumulated sufficient food.

Real ant colonies do not breed in this way, the workers being sterile, but we shall suspend realism for the sake of more interesting evolutionary programming.

7.3.5 Adding a Queen

Our ant simulation model has now become quite complex, so it is probably a good time to remind ourselves of the overall class structure (see Figure 7.6).

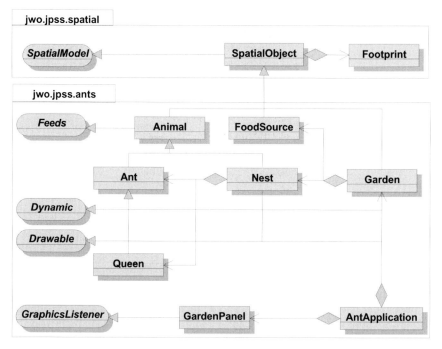

Figure 7.6 Ant class diagram with queen and breeding behaviour.

We have now created a garden containing potentially self-sustaining ant colonies. Ants can forage for food (which is continuously replenished), metabolise that food in order to continue to search, gather excess food and return it to its nest in order to produce new ants. We also hope that our ants will evolve over time to adapt their behaviour to the prevailing conditions. We will round off this case study by considering one final addition to our model that allows ants to build new nests (and therefore found new colonies) within the garden.

We will create a new type of ant – a Queen – that will behave much like a normal worker ant except that it will be able to move more quickly, and will never return to its original nest. That queen will feed just like any other ant, but unlike a worker, if it accumulates sufficient food, it will set up a new nest.

This behaviour is relatively simple to model with inheritance. We can create a new Queen class by extending the Ant class and modifying its evolve() method to account for queen-specific behaviour.

```
package jwo.jpss.ants;       // Part of the ant simulation package.
import java.awt.*;            // For drawing the queen ant.
import jwo.jpss.spatial.*;    // For spatial footprint.
import java.util.*;           // For dynamic collections.

// ************************************************************
/** Class for defining Queen ants.
  * @author Jo Wood
  * @version 1.4, 10th October, 2001.
  */
// ************************************************************

public class Queen extends Ant
{
    // ------------------- Class variables --------------------

                                  // Width of queen ant.
    private static final int QUEEN_WIDTH   = 15;
                                  // Height of queen ant.
    private static final int QUEEN_HEIGHT  = 9;
                                  // Rate at which ant consumes food.
    private static final int METABOLIC_RATE = 2;
                                  // Food per cycle that can be eaten.
    private static final int FEED_RATE     = 200;
                                  // Food level at which queen nests.
    private static final int NESTING_LEVEL = 20000;

    // -------------------- Constructors --------------------

    /** Creates a queen with similar characteristics of the given
      * ant.
      * @param foodLevel Initial food level of the ant.
      * @param parent Ant supplying inheritable characteristics.
      * @param x Initial x location of the ant.
      * @param y Initial y location of the ant.
      * @param nest Nest in which this ant was born.
      */
    public Queen(int foodLevel, Ant parent,
                 float x, float y, Nest nest)
    {
        super(foodLevel,parent,x,y,nest);

        // Make queen larger than worker ant.
```

```java
        setBounds(new Footprint(x,y,QUEEN_WIDTH,QUEEN_HEIGHT));

    // Queens are just like normal ants, but travel faster and
    // are less likely to move in circles.
    straightSteps *=2;
    changeAntDirection();
}

// ----------------- Overridden Methods ------------------

/** Draws the ant using the given graphics context.
  * @param g Graphics context to draw to.
  */
public void paint(Graphics g)
{
    Footprint fp = getBounds();
    int x = Math.round(fp.getXOrigin());
    int y = Math.round(fp.getYOrigin());
    int width = Math.round(fp.getMERWidth());
    int height = Math.round(fp.getMERHeight());

    g.setColor(colour);
    g.fillOval(x,y,width,height);
}

/** Let the queen go about its business for one time unit.
  */
public void evolve()
{
    metabolise(METABOLIC_RATE);

    if (isAlive())
    {
        // Find out if the ant is sitting on anything.
        Iterator i = gListener.objectsAt(this).iterator();
        while (i.hasNext())
        {
            SpatialObject spObject = (SpatialObject) i.next();

            // If queen is sitting on some food, eat it.
            if (spObject instanceof FoodSource)
            {
                // Attempt to eat food at feeding rate
                FoodSource foodSource = (FoodSource)spObject;
                int removed = foodSource.removeFood(FEED_RATE);

                if (removed > 0)
                {
                    eat(removed);
                    return;
                }
            }

            // If queen's food level is high, set up new nest.
            if ((spObject instanceof Garden) &&
                (getFoodLevel() >NESTING_LEVEL))
            {
                Garden garden = (Garden)spObject;
```

Collecting Objects Together

```
                    // Set straight steps back to normal ant level.
                    straightSteps /= 2;

                    // Add the new nest to the garden.
                    garden.requestNewNest(new Nest(this));

                    // Kill queen as has now become part of nest.
                    setAlive(false);
                }
            }

            // Change direction if any has walked sufficient steps.
            if (numSteps%straightSteps == 0)
               changeAntDirection();

            move(xDir,yDir);

            // Only move if within bounds.
            if (gListener.canDraw(this))
               numSteps++;
            else
            {
                move(-xDir,-yDir);
                changeAntDirection();
            }
        }
    }

    // --------------------- Private Methods ---------------------

    /** Sets a new direction for queen to walk in.
      */
    private void changeAntDirection()
    {
        // Queen chooses random direction and never returns home.
        xDir = (float)(Math.random()*10-5);
        yDir = (float)(Math.random()*10-5);
    }
}
```

Because much of the queen's behaviour is similar to a worker ant's, we do not need to repeat the code common to both. However, the evolve() method contains new nest building behaviour, so the Queen class *overrides* the evolve() method of the Ant class. Here, the queen feeds at its own FEED_RATE and if it has accumulated enough food, will create a new nest. In Section 7.3.3, we gave all ants sentience so they can identify all objects in their immediate surroundings. Since the Garden will be one of those objects, the queen can inform the garden of the new nest and its location. Once created, the nest will behave like any other, with new ants being produced that share the genes of the founding queen.

Figure 7.7 shows one possible evolution of the ant colonies when the simulation is run. The garden starts with two colonies and a large amount of food. Initially, ant production is likely to outpace the production of food, so after a short length of time, the garden is overrun by too many ants searching for too little food. Soon, many of the less successful ants start to die leaving fewer of the better fed ants in the garden. Occasionally, new queens emerge from the nest, and if they find sufficient food, they set up their own colonies.

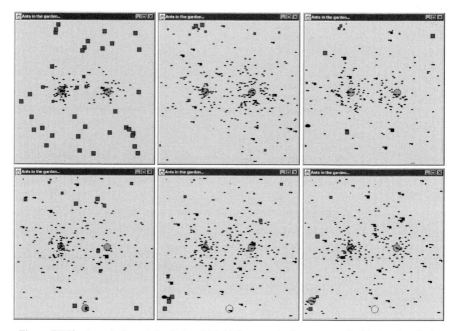

Figure 7.7 Six stages in the ant simulation: (a) initially two colonies are created with abundant food; (b) ant numbers increase with demand outstripping food supply; (c) ant numbers begin to dwindle and a new queen is produced (left of garden); (d) the queen sets up a new colony (bottom of garden); (e) there is not enough food to supply the new colony so it dies; and (f) eventually another queen creates a new colony and so the cycle continues.

7.4 SUMMARY

In this chapter, we have seen how we can group objects together using dynamic collections. These have considerable advantages over fixed size arrays as they allow items to be added and removed during the operation of a program. A number of different types of collection are available to the Java programmer depending on the way in which elements are organised.

Dynamic collections provide an ideal mechanism for representing spatial classes. We saw in this chapter how a GISVector class for storing coordinate pairs could be developed using the Java Vector collection. We also saw how a gazetteer could be created, which related names to spatial coordinates.

Adding dynamic collections to our ant simulation model allowed a garden to

Collecting Objects Together

contain a variable number of nests, which in turn can contain a dynamic collection of ants. This model was further developed to incorporate some ideas of evolutionary programming where ants gradually evolve to cope with their prevailing environmental conditions.

By the end of this chapter and associated exercises on the web, you should be able to

- understand how a GIS vector model represents the boundaries of a spatial object;
- create a dynamic collection using Java's `Collection` interface;
- determine which of Java's collections is most appropriate for modelling a given group of objects;
- use an iterator to cycle through the elements of a collection; and
- understand how genetic algorithms and evolutionary programming can be used to model certain processes.

CHAPTER EIGHT

Controlling Dynamic Events

Broadly speaking, the sequencing of the processes in our programs has been determined by the order in which lines of code appear in our Java programs. We have exercised quite sophisticated control over that order by creating and calling methods from a range of classes. We have passed information between methods and classes using parameter-based message passing. However, there comes a point in the development of many programs when we wish to respond to control at *runtime* rather than *compile-time*. In other words, the methods that are invoked are not determined by the programmer, but by the user of the program, or by external events that may vary over the time over which the program is run.

This chapter considers two ways in which we can account for dynamic runtime behaviour. The first is the use of *event handling*, associated largely with a graphical user interface's response to mouse clicks, button presses, etc. at runtime. The second is the division of dynamic processes into *threads* that may run in parallel to one another at runtime. Both allow the Java programmer to incorporate a greater degree of dynamism within their programs.

8.1 EVENT HANDLING

8.1.1 What is an Event?

When we create graphical interfaces in a windowing environment, even with a relatively simple windowing session, behind the scenes the computer is busy monitoring user input. In particular, the system has to keep track of each graphical component in case it is moved by the mouse, clicked on, resized etc. As most graphical programs will consist of many such components, that monitoring could be quite a complicated task.

Consequently, it is inappropriate to think of programming such a system in a linear fashion with a single start, middle and end. It is better to think of a set of objects that each listen out for dynamic processes such as mouse movements, button presses, etc. and act appropriately whenever these actions occur.

Such a dynamic action is known as an event, some examples of which include

- Menu item selected by the mouse
- A button is pressed
- Window is resized

- Window is iconised
- Some text is changed.

You can think of an event in much the same way as the object-oriented message – some form of communication between different objects. The only difference is that events are sent and received at runtime rather than being controlled by the programmer at compile-time. Most of the event handling you are initially likely to be exposed to are associated with some form of graphical process and often involve continuous streams of information being sent between objects.

The mechanism by which events are monitored and consequent actions are performed is known as an *event handler*.

8.1.2 Java Delegated Event Handling

Java uses an event handling process known as a *delegation event model*. In this model, the hierarchy of classes containing event handling routines are separated from the graphical components that generate them. If we wish a graphical component to respond to an event of some kind (such as being clicked on using the mouse), we need to tell it so by adding in the event listening code.

If you add a graphical component such as a JButton to a GUI, clicking on that button would have no effect. This is because so far we have not told the button component to 'listen' for mouse click events.

Consider the following program that adds the event handling code to a button so that it responds to being clicked:

```
import javax.swing.*;       // For Swing components
import java.awt.*;          // For AWT components.
import java.awt.event.*;    // For event listeners.

// ****************************************************************
/** Creates a simple Swing-based window and demonstrates how Java
 *  handles simple action events.
 *  @author    Jo Wood.
 *  @version   1.0, 17th June, 2001
 */
// ****************************************************************
public class Responsive extends JFrame implements ActionListener
{
    // -------------------- Object Variables --------------------

    private Container contentPane;   // This is the container into
                                     // which components are added.
    private JButton button;          // A button to press.

    // --------------------- Constructor ---------------------

    /** Creates a simple window with an equally simple label
```

```
            * inside.
            */
        public Responsive()
        {
            // Create the window with a title.
            super("Responsive Window");
            setDefaultCloseOperation(WindowConstants.DISPOSE_ON_CLOSE);
            contentPane = getContentPane();

            // Set the style of layout.
            contentPane.setLayout(new BorderLayout());

            // Create graphical components.
            JLabel label =new JLabel("A simple window", JLabel.CENTER);
            button = new JButton("Press Me");
            button.addActionListener(this);

            // Add graphical components to the window.
            contentPane.add(label,BorderLayout.NORTH);
            contentPane.add(button,BorderLayout.CENTER);

            // Get Java to size the window and make it visible.
            pack();
            setVisible(true);
        }

        // ---------------- Implemented methods ----------------

        /** Responds to a mouse click over the button.
         * @param event Mouse event.
         */
        public void actionPerformed(ActionEvent event)
        {
            if (event.getSource() == button)
            {
                contentPane.add(new JLabel("Ouch!",JLabel.CENTER),
                                BorderLayout.SOUTH);
                pack();
            }
        }
    }
```

When run, this produces a window similar to the one shown in Figure 8.1(a). When the button is pressed using the mouse the window enlarges and appears as Figure 8.1(b).

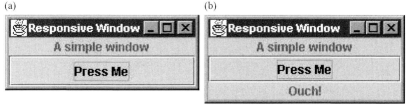

Figure 8.1 A simple window that responds to mouse click action events.

There are a number of new ideas contained within `Responsive`. First, note that since the event handling classes are stored in their own package, we need an extra import line at the top of the class definition (`import java.awt.event.*;`).

As in previous examples, to create our own window, we extend the `JFrame` class. Any container (like `JFrame`) holding a component that needs to respond to GUI actions such as button presses or menu selections should implement the `ActionListener` interface. This interface forces us to write a method called `actionPerformed()` which will include the code that responds to a button press event. The `actionPerformed()` method itself contains an object of type `ActionEvent` as an incoming message. This is the event that can be 'fired' whenever a button is pressed or a menu item is selected. Our method simply checks to see if it was the `button` object that was pressed, and adds a new label to the window if it was.

After initialising our button, we have to tell Java to listen for mouse presses:

`button.addActionListener(this);`

This is the code that delegates the event listening to the current class. In other words, it tells `button` to listen for mouse clicks and if such a click is heard, to go to the `actionPerformed()` method in `this` class. Note that because both the constructor and `actionPerformed()` need to know about the existence of `button`, it has been made an object variable rather than a local one (as both the labels were).

We can add event listeners to a range of components. The two most common circumstances when this is necessary are likely to be when you create buttons and menus. To consolidate, consider an example that creates a simple menu and listens for menu selection and window closing events.

```java
import javax.swing.*;        // For Swing components
import java.awt.*;           // For Dimension class.
import java.awt.event.*;     // For event listeners.

// ****************************************************************
/** Creates a simple Swing-based text editor window demonstrating
 *  how menus can be created and listened to.
 *  @author   Jo Wood.
 *  @version  1.1, 17th June, 2001
 */
// ****************************************************************

public class SimpleEditor extends JFrame implements ActionListener
{
    // -------------------- Object Variables --------------------

    private JMenuItem mExit;           // The 'Exit' menu item.

    // ---------------------- Constructor ----------------------

    /** Creates a window with a text editor inside.
```

```java
     */
    public SimpleEditor()
    {
        // Create the window with a title.
        super("Text Editor");
        setDefaultCloseOperation(
                            WindowConstants.DO_NOTHING_ON_CLOSE);
        addWindowListener(new WinMonitor());

        // Create Menu hierarchy.
        JMenuBar menuBar = new JMenuBar();
         JMenu menFile = new JMenu("File");
           mExit = new JMenuItem("Exit");
           mExit.addActionListener(this);
          menFile.add(mExit);
        menuBar.add(menFile);

        // Create editor window.
        JTextArea editWindow = new JTextArea("Some text to edit");
        JScrollPane scrollPane = new JScrollPane(editWindow);
        scrollPane.setPreferredSize(new Dimension(400, 120));

        // Add graphical components to the window.
        setJMenuBar(menuBar);
        getContentPane().add(scrollPane);

        // Get Java to size the window and make it visible.
        pack();
        setVisible(true);
    }

    // ----------------- Implemented methods -------------------

    /** Responds to a mouse click over a menu item.
      * @param event Mouse event.
      */
    public void actionPerformed(ActionEvent event)
    {
        if (event.getSource() == mExit)
            closeDown();
    }

    // ------------------- Private Methods ---------------------

    /** Asks the user if they really want to quit, then closes.
      */
    private void closeDown()
    {
        int response = JOptionPane.showConfirmDialog(this,
                            "Are you sure you want to quit?");
        if (response == JOptionPane.YES_OPTION)
            System.exit(0);    // Exit program.
    }

    // ------------------- Nested Classes ----------------------

    /** Monitors window closing events and performs a 'clean exit'
      * when requested.
      */
```

```
        private class WinMonitor extends WindowAdapter
        {
            /** Responds to attempt to close window via the GUI. Checks
             * the user really wants to quite before closing down.
             * @param event Window closing event.
             */
            public void windowClosing(WindowEvent event)
            {
                closeDown();
            }
        }
}
```

When an object is created out of this class, a window with active menus similar to that shown in Figure 8.2 is produced.

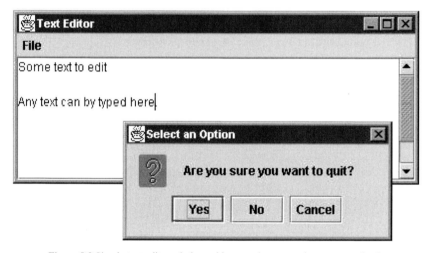

Figure 8.2 Simple text editor window with menu that responds to mouse selection.

The main text input area is a JTextArea that will allow text to be entered, edited, copied and pasted. This component is placed inside a JScrollPane in order to handle scrolling should the text inside be too large for the window in which it is placed.

The event handling code in this example is similar to the button event handling of the previous example. The only difference is that this time we have added the event listener to the mExit menu item rather than a button.

When creating menus in Java, we need to define three levels of object. The JMenuBar is the grey bar that holds all menus and is placed along the top of the window. The JMenu is the named menu that appears on the menu bar, while the JMenuItem is an individual item that appears on a menu when it is selected with the mouse.

Controlling Dynamic Events 199

SimpleEditor also uses event handling to provide a 'clean exit' from the application. If the user wishes to close the application, a window pops up asking the user if they are sure they want to quit (see Figure 8.2). If the user answers yes, the application exits. This is achieved by using the Swing component JOptionPane that carries out most of the work for us. This includes its own internal event handling that responds to a Yes/No/Cancel button press. Since this is encapsulated within the JOptionPane class, we need not worry how it is implemented.

It is possible that a user can also quit the application by clicking the close window symbol (a cross in the top-right corner of the window in Figure 8.2). We can add our own event handling that is informed when the user attempts to quit in this way. One way of doing this is to add a new event handler WindowListener to the JFrame. This can be done in much the same way as an ActionListener was added to buttons and menus.

```
public class SimpleEditor extends JFrame
                          implements ActionListener, WindowListener
{
    // ----------------- Constructor ----------------

    public SimpleEditor()
    {
        addWindowListener(this);

        // Other GUI code here.
    }
}
```

In this case, we add the WindowListener to the JFrame rather than JButton or JMenuItem. However, unlike the ActionListener interface, WindowListener forces us to implement several methods (see Table 8.1). Most of these methods are not needed by our SimpleEditor, so they must be created as 'empty' methods. For example,

```
/** Responds to a iconifying event but does nothing in this case.
  * @param event Window iconifying event (ignored).
  */
public void windowIconified(WindowEvent event)
{
    // Do nothing.
}
```

It is a little tedious and time-consuming for the programmer to add these seven methods to a class, only to make use of one of them. As a convenience, Java provides several *adapter classes* that already implement the interface with empty methods.

Table 8.1 Methods in the `WindowListener` interface

Method	Description
windowClosing(WindowEvent e)	Invoked when the user attempts to close the window via the GUI (usually a cross in the corner of the window).
windowActivated(WindowEvent e)	Invoked when the window becomes 'active' (i.e. the window is the one 'in focus', usually with a coloured bar across the top)
windowDeactivated(WindowEvent e)	Invoked when the window ceases to be 'active'. Usually because control has been transferred to some other window.
windowIconified(WindowEvent e)	Invoked when the window has been minimised to be replaced by an icon.
windowDeiconified(WindowEvent e)	Invoked if an iconified window is clicked, returning it to its normal status.
windowOpened(WindowEvent e)	Invoked when the window is made visible for the first time.
windowClosed(WindowEvent e)	Invoked when the window is finally closed (usually with the `dispose()` method)

All the programmer has to do is to subclass the adapter and override the method or methods they wish to use. Thus, if we wish to create a `WindowListener`, we can subclass `WindowAdapter`, and override the `windowClosing()` method.

```
private class WinMonitor extends WindowAdapter
{
    /** Responds to attempt to close window via the GUI. Checks
     * the user really wants to quite before closing down.
     * @param event Window closing event.
     */
    public void windowClosing(WindowEvent event)
    {
        closeDown();
    }
}
```

We delegate the event handling by creating an object from this class and informing the Window's listener:

`addWindowListener(new WinMonitor());`

`WinMonitor` can either be completely separate class, or it can be *nested* inside the class doing the delegating, as in the example above.

8.2 THREADS

We have moved a long way from the first Java programs introduced in Chapter 1. In our early programs, the sequencing of program instructions was simply a case of starting at the 'top' of a program, following instructions, line by line until the 'bottom' of the program was reached. The introduction of methods and looping structures complicated that pattern by allowing program control to jump within and between classes and to repeat instructions. Event handling took that process one stage further by allowing the user of a program rather than the programmer, to control the sequence in which lines of a program were executed.

Yet in all these examples, program execution has been essentially linear and serial. That is, one program line is acted upon before control moves to the next program instruction. A sequence of such instructions, even if it involves jumping between several methods and classes, is known as a *thread*. Java is known as a threaded language because it allows us control over the creation of such threads. Importantly, it allows us to create threads in *parallel* – multiple sequences of program control that run at the same time.

Why would we wish to create multiple threads in our programs? In many circumstances (as we have seen so far in this book), we can achieve our programming goals with a single sequence of program instructions. However, there are several circumstances when starting multiple parallel threaded processes can greatly improve the clarity and organisation of our programming. Examples include

- Speeding up the responsiveness of a graphical user interface by running the GUI and underlying processing code in parallel.
- Creating dynamic GUIs, such as animations.
- Allowing computationally intensive processes to run 'in the background' while other less intensive processes run in the 'foreground'.
- Modelling dynamic processes that respond to changes in real-time.

In Java, there are three ways in which multiple threads may be created – using a class that itself creates threaded processes; inheriting the `Thread` class; or implementing the `Runnable` interface. In this section, we will consider examples of all three approaches.

8.2.1 Creating a New Thread

One way in which we can set a task off in its own thread is to place that task inside a method inherited from the `Thread` class. The important method inside this class is `run()` – anything inside this method is kept in its own thread running along side any other threads in our program. For example, consider the following class that calculates whether or not a given number is prime (divisible only by itself and 1).

```
//   ****************************************************************
/** Determines if a given number is prime or not. The calculation
  * is executed as a threaded process.
  * @author Jo Wood.
  * @version 1.1, 27th October, 2001.
  */
```

```java
//   ****************************************************************
public class PrimeTest extends Thread
{
    // ------------- Object variables --------------

    private int n;          // Number to test.

    // --------------- Constructor ----------------

    /** Tests to see if the given number is prime.
      * @param n Number to test.
      */
    public PrimeTest(int n)
    {
        super();
        this.n = n;
    }

    // ------------- Overridden Methods ------------

    /** Looks for factors.
      */
    public void run()
    {
        for (int i=2; i<=n/2; i++)
        {
            if (n%i == 0)
            {
                System.out.println(n+" is not prime
                                    (divisible by "+i+")");
                return;    // Quit thread.
            }
        }
        System.out.println(n+" is prime");
    }
}
```

The calculation is straightforward, simply testing all integers from 2 to n/2, looking for factors. If none is found, the number is considered to be prime.

To start the thread, we must create an object out of the class and call its start() method. Thus, to test for several prime numbers, we can create multiple threads as follows:

```java
/** Tests for several prime numbers, each as a threaded process.
  */
public Prime()
{
    System.out.println("Starting calculation...");
    new PrimeTest(12345).start();
    new PrimeTest(70001).start();
    new PrimeTest(13).start();
    new PrimeTest(341).start();
    System.out.println("Haven't we finished yet?");
}
```

Unlike a conventional single-threaded program, these calculations will not be performed a 'top-down' order, but rather all threads will be started all at about the same time, and each will proceed with its calculation in parallel. Thus, output from the method above may be as follows:

```
Starting calculation...
Haven't we finished yet?
12345 is not prime (divisible by 3)
341 is not prime (divisible by 11)
13 is prime
70001 is prime
```

Results are reported in the order of time taken to complete the calculation. Thus, quick calculations (like determining that 12345 is divisible by 3) are reported more rapidly than longer ones (like determining that 70001 has no factors). Notice also that the 'Haven't we finished yet?' message is displayed before any of the threads have completed their calculations, despite being invoked after all threads have been started. On a faster computer, it is possible that some of the threads might complete before this text is displayed. This introduces one of the complications with threaded, or what tends to be known as *concurrent programming*. Unlike simple sequential programming, it is often not possible to predict with any certainty, the exact sequencing of program execution. Once multiple threads are running, the time taken to complete their respective tasks can vary from machine to machine, or even between successive invocations on the same machine.

A second problem we face with extending the Thread class is that we cannot also inherit any other classes at the same time. This is because Java does not allow *multiple inheritance*. We can however, *implement* multiple interfaces in our classes. So, an alternative way of creating a threaded process is to implement the Runnable interface. This forces the class to implement a run() method and allows us to define a start() method that creates a threaded process. So, to adapt PrimeTest to use a Runnable interface, we can do the following:

```
public class PrimeTest implements Runnable
{
    /** Tests to see if the given number is prime.
      * @param n Number to test.
      */
    public PrimeTest(int n)
    {
        // Constructor here.
    }

    /** Starts the threaded process.
      */
    public void start()
    {
        Thread thread = new Thread(this);
        thread.start();
    }
```

```
        /** Looks for factors.
         */
        public void run()
        {
            // Calculation here.
        }
}
```

The line `Thread thread = new Thread(this);` creates the new thread, and passes to its constructor any class that has a `run()` method. Since `PrimeTest` implements `Runnable`, passing `this` as an argument is sufficient to tell the thread to use `PrimeTest`'s `run()` method to perform the threaded processing.

As you can see, implementing threaded processes using `Runnable` is slightly more complicated than inheriting `Thread`, but it has the significant advantage of allowing a class to inherit something other than `Thread` yet still perform processing in parallel with other operations.

An alternative to both inheriting `Thread` and implementing `Runnable` involves using a higher level class that itself is threaded. Java includes many threaded classes, such as those required to create GUIs. In this section, we will consider a simple class `Timer` (part of the `java.util` package) that can be used to keep track of time 'in the background' and schedule method calls at regularly-timed intervals.

The example below uses `Timer` to create a simple quiz that asks the user to answer a question within a certain time limit.

```
import java.util.*;              // For the threaded timer.
import jwo.jpss.utilities.*;     // For keyboard input.
//   ****************************************************************
/** Creates timed quiz questions. Demonstrates the use of the
  * threaded timer to control dynamic processes.
  * @author Jo Wood.
  * @version  1.1, 27th October, 2001
  */
//   ****************************************************************

public class Quiz
{
    // --------------- Object and class variables ----------------

    private Timer timer;
    private boolean timesUp;
    private static final int TIME_LIMIT=5;   // Time in seconds.


    // --------------------- Constructor ----------------------

    /** Asks a simple timed question
      */
    public Quiz()
```

```java
    {
        // Initialise questions, answers and timer.
        KeyboardInput keyIn = new KeyboardInput();
        timer = new Timer();
        timesUp = false;
        String questions[]={"What do the initials GPS stand for?"};
        String answers[]   ={"Global Positioning System"};

        // Asks question.
        System.out.println("You have "+TIME_LIMIT+
                   " seconds to answer the following question...");

        timer.schedule(new Countdown(),TIME_LIMIT*1000);
        keyIn.prompt(questions[0]);
        timer.cancel();

        // Give feedback.
        if (timesUp)
            System.out.println("Sorry - you took too long.");
        else
            if (keyIn.getString().equalsIgnoreCase(answers[0]))
                System.out.println("Well done!");
            else
                System.out.println("No, the answer is "+answers[0]);
    }

    // -------------------- Nested Classes ---------------------

    /** Indicates that a scheduled time limit has been reached.
      */
    private class Countdown extends TimerTask
    {
        /** Indicates that time is up.
          */
        public void run()
        {
            System.out.println("\nPing!");
            timesUp = true;
            timer.cancel();
        }
    }
}
```

`Timer` is a threaded class that keeps track of elapsed time in the background. The `schedule()` method informs the class that the `run()` method of a `TimerTask` should be called after a given time interval (5 seconds in this example). In the meantime, `Quiz`'s constructor asks a question and analyses the response from the user. If this can be done before the `run()` method is called, the question is 'marked', if not, the user is informed that they have run out of time.

Following program flow is a little more complicated once multiple processes are introduced. In this case, it is further complicated by the fact that one of the processes operates in real-time. The sequencing of events is more clearly illustrated in Figure 8.3, showing the interaction of the two threaded processes.

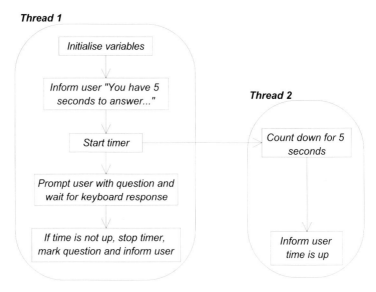

Figure 8.3 The sequence of two threaded processes in the Quiz class.

Threads and Parallel Processing

On most systems, multiple threads do not actually run genuinely in parallel. Single processor computers (which include most PCs), can only process one low-level instruction at a time. Multithreading only gives the *illusion* of parallel processing, so should be considered as a way of making programs clearer rather than necessarily more efficient.

Multiprocessor computers can handle parallel threads on different processors, but require a Java Virtual Machine to allocate threads appropriately.

8.2.2 Using Threads with a Graphical User Interface

One of the commonest uses of threaded processes is to ensure that a graphical user interface remains responsive to user input. In fact, many of the GUI-based programs, we have already created use threaded processes to handle, apparently simultaneously, the many possible demands put on the GUI such as drawing windows, resizing windows, listening for mouse interactions, etc. Thankfully, for us as programmers, many of the detailed workings of the GUI are encapsulated within the AWT and Swing classes.

We are more likely to require explicit multithreading when we develop a GUI for some application that involves more demanding computational analysis. In such cases, we want to be able to allocate resources both to the computation and to ensuring that the GUI still responds to window resizes, event listening, etc. We would like to be able to avoid the GUI apparently 'locking up' while the computer devotes all its resources to performing some calculation.

The easiest way of avoiding such problems is to build a GUI as normal, but to assign any 'computationally expensive' tasks to their own threads. The example below shows how we might build a GUI around the prime number class developed previously. The result of running the program (see Figure 8.4) is a window that allows any number to be typed in. Factors of that number are searched for in their own thread with a 'progress bar' indicating processing. The application allows many numbers to be calculated simultaneously, as each is allocated its own thread. Even when several numbers are being calculated, the program remains responsive to mouse and keyboard input, allowing the window to be resized or quit.

```java
import javax.swing.*;         // For GUI components.
import java.awt.*;
import java.awt.event.*;      // For event handling.

// ****************************************************************
/** Graphical User Interface for exploring prime numbers.
  * @author Jo Wood.
  * @version 1.2, 27th October, 2001.
  */
// ****************************************************************

public class PrimeGUI extends JFrame implements ActionListener
{
    // -------------------- Object Variables ---------------------

    private JButton bCalc;
    private JTextField tfInput;
    private JTextArea taOutput;
    private JPanel pStatus;

    // --------------------- Constructor -----------------------

    /** Creates a GUI for testing prime numbers.
      */
    public PrimeGUI()
    {
        super("Prime Number Explorer");
        setDefaultCloseOperation(
                          WindowConstants.DO_NOTHING_ON_CLOSE);
        addWindowListener(new WinMonitor());
        Container contentPane = getContentPane();

        // Create input area.
        JPanel pInput = new JPanel();
        pInput.add(new JLabel("Type in a number: "));
        tfInput = new JTextField(8);
        pInput.add(tfInput);
        bCalc = new JButton("Calculate");
        bCalc.addActionListener(this);
        pInput.add(bCalc);
```

```java
        contentPane.add(pInput, BorderLayout.NORTH);

        // Create output area.
        taOutput = new JTextArea();
        taOutput.setEditable(false);
        JScrollPane scrollPane = new JScrollPane(taOutput);
        scrollPane.setPreferredSize(new Dimension(300,200));
        contentPane.add(scrollPane, BorderLayout.CENTER);

        // Create status bar.
        pStatus = new JPanel();
        contentPane.add(pStatus, BorderLayout.SOUTH);

        // Size and show window.
        pack();
        setVisible(true);
    }

    /** Responds to a mouse click over the button.
      * @param event Mouse event.
      */
    public void actionPerformed(ActionEvent event)
    {
        if (event.getSource() == bCalc)
        {
            int n = Integer.parseInt(tfInput.getText());
            new PrimeTest(n).start();
        }
    }

    // -------------------- Private Methods ----------------------

    /** Asks the user if they really want to quit, then closes.
      */
    private void closeDown()
    {
        int response = JOptionPane.showConfirmDialog(this,
                            "Are you sure you want to quit?");
        if (response == JOptionPane.YES_OPTION)
            System.exit(0);    // Exit program.
    }

    // -------------------- Nested Classes ----------------------

    /** Determines if a given number is prime or not.
      */
    private class PrimeTest extends Thread
    {
        // ------------- Object variables --------------

        private int n;       // Number to test.

        // -------------- Constructor ----------------

        /** Tests to see if the given number is prime.
          * @param n Number to test.
          */
        public PrimeTest(int n)
        {
```

```
            super();
            this.n = n;
        }

        /** Looks for factors.
          */
        public void run()
        {
            // Set up progress bar.
            JProgressBar progress = new JProgressBar(2,n/2);
            progress.setString(n+" ");
            progress.setStringPainted(true);
            pStatus.add(progress);
            pack();

            // Search for factors.
            for (int i=2; i<=n/2; i++)
            {
                progress.setValue(i);
                if (n%i == 0)
                {
                    taOutput.append(n+
                            " is not prime (divisible by "+i+")\n");
                    pStatus.remove(progress);
                    pack();
                    return;    // Quit thread.
                }
            }
            taOutput.append(n+" is prime\n");
            pStatus.remove(progress);
            pack();
        }
    }

    /** Monitors window closing events and performs a 'clean exit'
      * when requested.
      */
    private class WinMonitor extends WindowAdapter
    {
        /** Checks user really wants to quite before closing down.
          * @param event Window closing event.
          */
        public void windowClosing(WindowEvent event)
        {
            closeDown();
        }
    }
}
```

The only new aspect of this code is the use of the Swing class `JProgressBar`. The constructor of this class takes a start and end value as parameters. Any number supplied to its method `setValue()` will shade the progress bar in proportion to the range initialised by the constructor. Its `setString()` method is used to display the number being analysed while progress is shown. Every time a new progress bar is added (by the `PrimeTest` thread), the window is 'repacked' allowing it to resize to fit the necessary number of progress bars along the bottom.

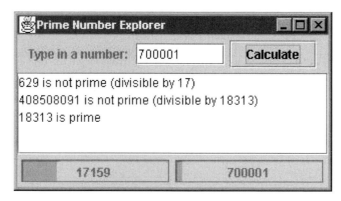

Figure 8.4 Output from the Prime Number Explorer. The progress bars at the bottom of the screen indicate processing of outstanding numbers.

8.3 CASE STUDY: CREATING A DISPLAYABLE SPATIAL MODEL

In this case study, we will take the `RasterMap` and `VectorMap` models created in Chapters 4 and 7, and create a graphical user interface for displaying both. This process uses some of the techniques described in Chapter 4 for displaying the raster model, but uses object-oriented modelling to generalise the process allowing both rasters and vectors to be displayed. It will add various event listeners to the GUI so that it can keep track of the size of the display and perform simple mouse-based query operations.

8.3.1 Improving the Class Design

We have developed a number of classes in previous chapters that can be used for representing different aspects of spatial information. These are shown in Table 8.2.

Table 8.2 Spatial classes and interfaces previously developed

Class	Description
RasterMap	Represents images and regular grids of spatial data.
GISVector	Represents a spatial object by its geometric boundary.
VectorMap	A collection of `GISVector`s.
SpatialObject	A generic class for representing spatial things (e.g. Ants).
Footprint	Stores the geometry of a spatial location.
Gazetteer	Attaches names to spatial locations.
SpatialModel	Interface describing the types of behaviour all spatial models should have.

It makes sense to group all of these classes together and where possible, force them to share common functionality. To a large extent, we have already done this and in Chapter 6, we created a package jwo.jpss.spatial, into which such classes can be placed. It is a relatively easy task to take the RasterMap developed in Chapter 4, force it to subclass SpatialObject, and place it in the jwo.jpss.spatial package. The overall class design of that package is shown in Figure 8.5.

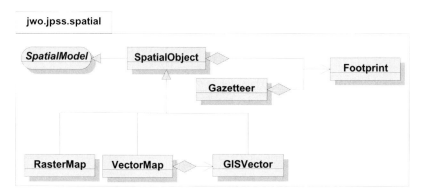

Figure 8.5 Classes and interfaces in the jwo.jpss.spatial package.

Two aspects are lacking from the class design as it stands. Currently, there is no mechanism for storing extra textual information such as title, owner and copyright details associated with the spatial models. Secondly, it would be useful to add a GUI to display all spatial objects.

Storing extra textual information is straightforwardly achieved by creating a new class, Header with relevant accessor and mutator methods.

```
package jwo.jpss.spatial; // Part of the spatial modelling package.

//    ****************************************************************
/** Stores general header information about a document. Includes
  * information such as title, author, copyright details etc.
  * @author Jo Wood.
  * @version 2.0, 6th September, 2001.
  */
//    ****************************************************************
public class Header
{
    // ------------------ Object variables ---------------------

    private String title, author, rights, notes;

    // ------------------- Constructors ----------------------

    /** Creates a default header with a given title.
      * @param title Title of object described by the header.
      */
```

```java
   public Header(String title)
   {
      this (title, "","","");
   }

   /** Creates header with given title, author, rights and notes.
     * @param title Title of object described by the header.
     * @param author Author of object described by the header.
     * @param rights Copyright details associated with object.
     * @param notes Miscellaneous notes associated with object.
     */
   public Header(String title, String author,
                 String rights, String notes)
   {
      this.title  = title;
      this.author = author;
      this.rights = rights;
      this.notes  = notes;
   }
// -------------------- Accessor Methods --------------------

   /** Reports title of the object associated with this header.
     * @return Title of object.
     */
   public String getTitle()
   {
      return title;
   }

   /** Reports author of the object associated with this header.
     * @return Author of object.
     */
   public String getAuthor()
   {
      return author;
   }

   /** Reports the copyright details of the object associated
     * with this header.
     * @return Copyright details associated with this object.
     */
   public String getRights()
   {
      return rights;
   }

   /** Reports notes on the object associated with this header.
     * @return Notes associated with object.
     */
   public String getNotes()
   {
      return notes;
   }

   /** Reports the contents of the header as a text string.
     * @return Text representing header.
     */
   public String toString()
   {
```

Controlling Dynamic Events

```
            return new String(title +" "+ notes);
        }
        // ------------------- Mutator Methods --------------------

        /** Sets the title of the object associated with this header.
          * @param title New title of object.
          */
        public void setTitle(String title)
        {
            this.title = title;
        }

        /** Sets the author of the object associated with this header.
          * @param title New author information associated with object.
          */
        public void setAuthor(String author)
        {
            this.author = author;
        }

        /** Sets the copyright details associated with this header.
          * @param rights Copyright information associated with object.
          */
        public void setRights(String rights)
        {
            this.rights = rights;
        }

        /** Sets the notes on the object associated with this header.
          * @param notes New notes associated with object.
          */
        public void setNotes(String notes)
        {
            this.notes = notes;
        }
}
```

As any spatial object might be associated with such a header, it makes sense to add an accessor method to the `SpatialModel` interface that reports the header of a spatial model.

Accessor method in the `SpatialModel` interface

```
    /** Reports the header information associated with this object.
      * @return Header information (title, copyright etc).
      */
    public Header getHeader();
```

We can add these text headers to any of the spatial models by passing them as messages to the constructor of each. So for example, the constructors of `RasterMap` are shown below.

RasterMap *constructors*

```
/** Creates a raster map with the given dimensions.
  * @param numRows Number of rows in raster.
  * @param numCols Number of columns in raster.
  * @param res Size of 1 side of a raster cell.
  * @param fp Location of bottom left corner of raster.
  */
public RasterMap(int numRows, int numCols, float res, Footprint fp)
{
    this(numRows,numCols,res,fp, new Header("Simple raster"));
}

/** Creates a raster map with the given dimensions.
  * @param numRows Number of rows in raster.
  * @param numCols Number of columns in raster.
  * @param res Size of 1 side of a raster cell.
  * @param fp Location of bottom left corner of raster.
  * @param header Header associated with this raster.
  */
public RasterMap(int numRows, int numCols,
                 float res, Footprint fp, Header header)
{
    // Store incoming messages as object variables.
    this.numRows    = numRows;
    this.numCols    = numCols;
    this.resolution = res;
    setBounds(new Footprint(fp.getXOrigin(),fp.getYOrigin(),
                            res*numCols,res*numRows));
    setHeader(header);

    // Declare and initialise new raster array.
    raster = new float[numRows][numCols];

    for (int row=0; row<numRows; row++)
        for (int col=0; col<numCols; col++)
            raster[row][col] = 0.0f;
}
```

Our second activity to improve the overall class design of the spatial package is to incorporate a graphical user interface that allows spatial models to be displayed and queried graphically. As was suggested in earlier chapters, keeping the GUI development separate from the data structures holding and processing the spatial data keeps the design as flexible as possible. For this reason, we shall develop a separate sub-package for storing all the GUI classes that display the spatial models (jwo.jpss.spatial.gui).

You may recall from Chapter 4, that when we designed a GUI to display raster maps, we created a special panel onto which the contents of a RasterMap were displayed. This panel was placed in a JFrame along with textual metadata associated with the raster. We can generalise this approach by creating a new class SpatialPanel for displaying any spatial object and a container class ModelDisplay for showing the SpatialPanel and the textual data held in the Header class. We will need to subclass SpatialPanel to deal with any drawing code specific to either `RasterMaps` or `VectorMaps`. The overall class design of the package is shown in Figure 8.6.

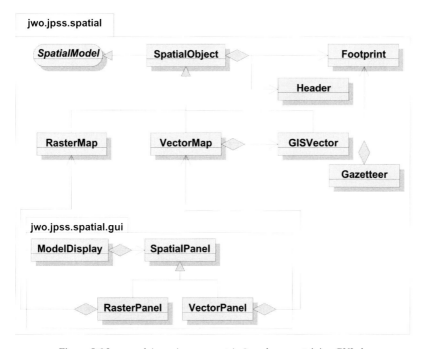

Figure 8.6 Improved `jwo.jpss.spatial` package containing GUI classes.

8.3.2 Creating a responsive `SpatialPanel`

As we did for the original design for `RasterDisplay` in Chapter 4, it is useful to sketch out the design of a GUI before coding it. The intention is to produce a similar GUI to the raster display, but one that can be generalised to display any spatial model. We would also like the GUI to respond to simple mouse-based query of the spatial model. Figure 8.7 shows a preliminary design.

The `JFrame` contains a `SpatialPanel` on the left showing a graphical representation of the spatial model (whether a `RasterMap`, `VectorMap` or some as yet undefined `SpatialObject`). This panel needs to redraw itself whenever the window is resized, scaling the objects within it appropriately. We would also like to be able to query the objects drawn on this panel using the mouse. The results of the query should be displayed in the 'status bar' along the bottom of the frame. Rescaling the graphics and listening for mouse queries both involve event handling in a similar way to the button presses and menu selections seen earlier in this chapter. But before we examine the event handling, we will consider the overall structure of the `SpatialPanel` class.

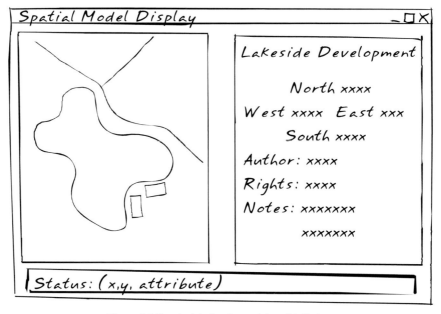

Figure 8.7 Sketched design for spatial model display.

`SpatialPanel` needs to contain the GUI functionality common to all types of `SpatialModel`. The class can then be inherited by any particular spatial model (for example, a `VectorMap`), and code specific to that form of model can be added. The general requirements of a `SpatialModel` are:

- to scale the spatial model so it fits within the boundaries of the panel;
- to listen for mouse input over the area of the panel;
- to translate any mouse input into its equivalent georeferenced coordinates;
- to rescale the spatial model and mouse input if the panel is ever resized; and
- to be able to send text output to other components (e.g. the status bar).

The drawing of the model can then be left to a subclass as this will vary depending on the type of model.

Event Handling

Three forms of event handling are necessary from the requirements list above, namely `listening` for window resizing, for mouse movement over the panel and for mouse clicking over the panel. In Java, these are handled by the `ComponentListener`, `MouseMotionListener` and `MouseListener` respectively.

A `ComponentListener` keeps track of changes in size, location and visibility of any graphical component. Normally, we are not interested in any of these

Controlling Dynamic Events 217

changes since Java's window handling and layout manager does all the work for us. In our particular example though, we need to know when the panel changes size so that we can rescale the graphics to fit inside the resized panel.

Mouse listening is handled by two separate listeners. `MouseListener` keeps an ear out for mouse presses and stores the location (in screen pixel coordinates) of the mouse at the point at which the mouse was pressed, dragged and released. In our example, we wish to report the georeferenced location and attribute at the point on our panel at which the mouse has been pressed. A separate listener, `MouseMotionListener`, can also keep track of the current mouse position as it is moved over a component. We will use this to report continuously, the georeferenced location represented by the current mouse position.

All three listener interfaces contain methods that we do not need to use (for example, responding to the mouse being dragged), so it makes sense to use nested *adapter methods* as we did in Section 8.1.2. Therefore, the overall structure of `SpatialPanel` along with its event handling code looks as follows:

```
package jwo.jpss.spatial.gui;   // Part of the spatial GUI package.

import jwo.jpss.spatial.*;      // For spatial classes.
import javax.swing.*;           // For Swing components.
import java.awt.*;              // For cursor handling.
import java.awt.event.*;        // For mouse events.

// ****************************************************************
/** Panel for displaying spatial objects.
  * @author Jo Wood.
  * @version 1.1, 21st October, 2001.
  */
// ****************************************************************

public class SpatialPanel extends JPanel
{
    // --------------- Object and Class Variables ----------------

    private SpatialModel spModel;       // Spatial model to display.

    // --------------------- Constructor ----------------------

    /** Creates a panel for displaying the given spatial model.
      * @param spModel Spatial model to display.
      */
    public SpatialPanel(SpatialModel)
    {
        super();
        this.spModel = spModel;

        setCursor(new Cursor(Cursor.CROSSHAIR_CURSOR));
        addMouseListener(new MouseClickMonitor());
        addMouseMotionListener(new MouseMoveMonitor());
        addComponentListener(new PanelSizeMonitor());
    }

    // -------------------- Nested Classes ----------------------
```

```java
    /** Handles mouse clicks on the panel.
     */
    private class MouseClickMonitor extends MouseAdapter
    {
        /** Handles a mouse click in the panel and reports the
         * coordinates of the mouse when clicked.
         * @param e Mouse event associated with the click.
         */
        public void mousePressed(MouseEvent e)
        {
            // Report location and attribute here.
        }
    }

    /** Handles mouse movement over the panel.
     */
    private class MouseMoveMonitor extends MouseMotionAdapter
    {
        /** Handles mouse movement over the panel by reporting
         * the coordinates of the current mouse position.
         * @param e Mouse event associated with movement.
         */
        public void mouseMoved(MouseEvent e)
        {
            // Report location here.
        }
    }

    /** Handles changes in the panel's status.
     */
    private class PanelSizeMonitor extends ComponentAdapter
    {
        /** Handles panel resizing by updating the pixel-
         * georeferenced transformation to account for new size.
         * @param e Panel resizing event.
         */
        public void componentResized(ComponentEvent e)
        {
            // Calculate pixel-georeferenced transformation here.
        }
    }
}
```

The only new Java code here is the use of the setCursor method that comes with all Components. This changes the default cursor into a cross-hair design whenever the mouse is over the panel.

Spatial Transformations

The main job of the SpatialPanel is to provide a mechanism to convert between the pixel coordinates used for display and mouse input and the georeferenced coordinates used by our spatial models (see Figure 8.8). The transformation needs to work in both 'directions' as we will need to convert mouse input coordinates in pixels into georeferenced coordinates as well as georeferenced coordinates back into pixels for drawing.

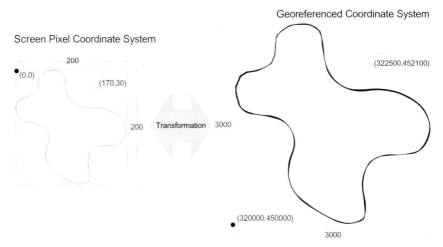

Figure 8.8 Pixel and georeferenced coordinate systems.

The types coordinate transformations required to do this are known as *affine transformations*. An affine transformation allows a geometric object to be translated (moved to a new location), scaled (made larger, smaller or stretched) and rotated. In our example, to change screen coordinates into georeferenced ones, we will need to

- *translate* the screen origin (0, 0) to the georeferenced origin;
- flip the Y axis so it increases upwards from the bottom of the coordinate space (this is actually a *scaling* of the y coordinates by –1); and
- *scale* the screen coordinates to use the georeferenced units.

We will also need to provide a similar set of transformations to convert georeferenced coordinates back into screen coordinates.

One of the useful properties of an affine transformation is that successive affine transformations (as we are required to perform above) can be combined in a single stage. The (matrix) mathematics required to do this need not bother us at the moment, as Java provides a class, AffineTransform, to do the work for us. This class is part of the java.awt.geom package, so it needs to be imported explicitly at the top of any classes that use it.

We can create a method in SpatialPanel to calculate the required transformations so that whenever we need to convert between screen and georeferenced coordinates, we can use the relevant affine transformation.

```
/** Calculates the transformations required to convert between
  * pixel coordinates and georeferenced coordinates.
  */
public void calcTransformation()
{
```

```
        // Scale spatial bounds to fit inside panel and flip Y axis.
        int panelWidth  = getWidth()  - 2*BORDER_SIZE;
        int panelHeight = getHeight() - 2*BORDER_SIZE;

        trans = new AffineTransform();
        iTrans = new AffineTransform();

        Footprint fp = spModel.getBounds();
        double scaling = Math.min(panelWidth/fp.getMERWidth(),
                                  panelHeight/fp.getMERHeight());

        float centreX = (panelWidth  - (fp.getMERWidth()*scaling))/2;
        float centreY = (panelHeight - (fp.getMERHeight()*scaling))/-2;

        trans.translate(BORDER_SIZE+centreX,
                        panelHeight+BORDER_SIZE+centreY);
        trans.scale(scaling,-scaling);
        trans.translate(-fp.getXOrigin(), -fp.getYOrigin());

        iTrans.translate(fp.getXOrigin(),fp.getYOrigin());
        iTrans.scale(1/scaling,-1/scaling);
        iTrans.translate(-centreX-BORDER_SIZE,
                         -centreY-panelHeight-BORDER_SIZE);
}
```

This method calculates the georeferenced to screen transformation (`trans`) by finding the screen dimensions of the panel and the georeferenced dimensions of the spatial object. Note that we have allowed for an internal border within the panel defined by the static constants `BORDER_SIZE`. If the bounds of the spatial object or the panel are not square, there will be a gap either above and below the feature, or to the left and right. The variables `centreX` and `centreY` calculate the translation required to make sure that any gaps are equally spaced around the edges.

The transformation itself is calculated by successive calls to methods in the `AffineTransform` class (`scale()`, and `translate()` are used here, but the class also contains an equivalent `rotate()` method). Finally, the inverse transformation from screen to georeferenced coordinates is also calculated.

The only times this method needs to be called are when either the size of the panel or the spatial object is changed. The event handling routines discussed above can be used to update the transformation if the user ever causes the panel to be resized.

So far, we have calculated and stored the transformations required to convert between coordinate systems. What we have not yet coded, is anything that actually performs the necessary transformations. To do this, we can create two new methods that receive as a message a location in one coordinate system and return the equivalent location in the other system.

```
/** Transforms given footprint from pixel to georeferenced coords.
  * @param fp Pixel coordinates to transform.
  * @return Georeferened coordinates of the given footprint.
```

Controlling Dynamic Events

```
     */
    public Footprint getPixelToGeo(Footprint fp)
    {
        Point2D geo,pxl,geoMax,pxlMax;

        geo = new Point2D.Float(fp.getXOrigin(),fp.getYOrigin());
        pxl = new Point2D.Float();
        iTrans.transform(geo,pxl);

        if ((fp.getMERWidth()==0) && (fp.getMERHeight()==0))
            return new Footprint((float)pxl.getX(),(float)pxl.getY());
        else
        {
            geoMax=new Point2D.Float(fp.getXOrigin()+fp.getMERWidth(),
                                     fp.getYOrigin()+fp.getMERHeight());
            pxlMax = new Point2D.Float();
            iTrans.transform(geoMax,pxlMax);
            return new Footprint((float)pxl.getX(),(float)pxl.getY(),
                                 (float)(pxlMax.getX()-pxl.getX()),
                                 (float)(pxlMax.getY()-pxl.getY()));
        }
    }
}
```

This method uses a new Java class called Point2D to store 2D locations. This class is part of the java.awt.geom package along with the AffineTransform class. The method takes the pixel coordinates from an incoming Footprint message, places them in a Point2D object and performs the affine transformation calculated previously. The if-else construction allows the transformation to be applied either to a single point, or to a rectangular area stored in the Footprint.

An equivalent method (getGeoToPixel()) performs the transformation in the other direction, but uses trans rather than iTrans to transform the coordinates.

8.3.3 Displaying Vector Maps

With the creation of the SpatialPanel class, we are free to inherit it in order to add code specifically to draw VectorMaps. To do this, we will make extensive use of what is known as the Java2D API. In fact, we have already been using this as the AffineTransform and Point2D classes are both part of Java2D.

Java2D offers considerable control over the way in which graphics are displayed in GUIs. It is particularly suitable for displaying both raster and vector images. The classes in the Java2D packages are also useful to the spatial programmer as they contain functionality for a range of spatial operations such as point-in-polygon and line intersection testing.

Implementing a class that uses Java2D is pretty straightforward as it follows the same structure as our other customised graphics classes. We simply inherit an existing container class (SpatialPanel in this example), and override its paintComponent() method.

```java
package jwo.jpss.spatial.gui;   // Part of the spatial GUI package.

import jwo.jpss.spatial.*;      // For spatial classes.
import java.awt.*;              // For graphics context.
import java.awt.geom.*;         // For 2D geometry display.
import java.util.*;             // For dynamic collections.

// ****************************************************************
/** Panel that draws a GIS vector map.
  * @author Jo Wood.
  * @version 1.3, 21st October, 2001
  */
// ****************************************************************

public class VectorPanel extends SpatialPanel
{
    // ---------------- Object and Class variables ---------------

    private VectorMap vectorMap;        // Vector map to draw.

    // Drawing styles.
    private static final int LINE_WIDTH = 2;
    private static final BasicStroke
                       STROKE = new BasicStroke(LINE_WIDTH);
    private static final Color
                       BACKGROUND_COLOUR - new Color(250,250,240);
    private static final Color
                       BORDER_COLOUR = new Color(0,0,0);
    private static final Color
                       FILL_COLOUR = new Color(240,240,180,100);


    // --------------------- Constructor ----------------------

    /** Creates a panel and draws the given vector map upon it.
      * @param vectorMap Vector map to draw.
      */
    public VectorPanel(VectorMap vectorMap)
    {
        super(vectorMap);
        setBackground(new Color(255,255,255,0)); // Transparent.
        this.vectorMap = vectorMap;
    }

    // --------------------- Methods --------------------------

    /** Draws graphics on the panel.
      * @param g Graphics context in which to draw.
      */
    public void paintComponent(Graphics g)
    {
        super.paintComponent(g);        // Paint background.
        Graphics2D g2 = (Graphics2D) g; // Use Java2D graphics
        g2.setRenderingHint(RenderingHints.KEY_ANTIALIASING,
                            RenderingHints.VALUE_ANTIALIAS_ON);

        // Draw each of the GISVectors
        Iterator i = vectorMap.getGISVectors().iterator();
```

```
            while (i.hasNext())
            {
                GISVector gisVect = (GISVector)i.next();
                GeneralPath path = gisVect.getCoords();
                path.transform(getGeoToPixel());

                if (gisVect.getType() == SpatialModel.AREA)
                {
                    g2.setPaint(FILL_COLOUR);
                    g2.fill(path);
                }

                g2.setPaint(BORDER_COLOUR);
                g2.setStroke(STROKE);
                g2.draw(path);

                // Restore vector object back to georeferenced coords.
                path.transform(getPixelToGeo());
            }
        }
    }
```

Unlike the previous examples of graphics customisation, the `paintComponent()` method typecasts the graphics context (`Graphics`) into its Java2D equivalent (`Graphics2D`). This releases a new set of methods available to us to draw high-quality vector graphics.

Of particular interest are the methods `setPaint()`, `setStroke()`, `draw()` and `fill()`. Respectively, these methods determine the colour in which objects are drawn, the style in which they are drawn, draws an object's outline, and draws an object as a filled area.

We can control the width and style of any lines drawn by creating a `Stroke` object. In the example above, this is achieved by creating a static constant with a fixed two pixel wide line width. The appearance of all lines and areas drawn using Java2D can also be controlled by setting what are called *rendering hints*. These can be thought of as advice given to the Java runtime environment as to how you as the programmer would like things drawn. Java is free to ignore this advice if, for example, the local hardware is not capable of drawing in the requested style. Figure 8.9 shows a selection of possible rendering styles.

The coordinates that make up a `GISVector`'s boundaries have been placed in the Java2D structure `GeneralPath`. This can be thought of as Java's equivalent to our own `GISVector` object. It stores a set of coordinate pairs, but conveniently for us, these coordinates can be transformed and drawn by supplying them as parameters to the relevant methods (`trans()` and `draw()` above). So, how did the GIS vector coordinates become represented as a `GeneralPath`? The answer lies in the modification made to the `GISVector` class that allows it to return the vector coordinates in the correct form.

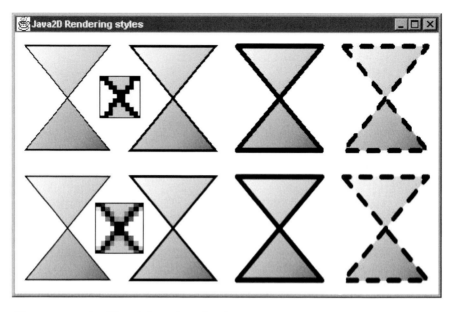

Figure 8.9 Some Java2D rendering styles and strokes. Stroke widths of 1, 2, 4 and 4 from left to right. Final column shows a dashed stroke. Upper row shows default rendering style, the lower row shows anti-aliasing of strokes (inset shows detail at intersection of two lines).

The original design of `GISVector` developed in Chapter 7 used a `Vector` of `Footprints` to store coordinates. The following shows the modified `GISVector` class with selected methods that demonstrate how `GeneralPath` has been used instead.

```
package jwo.jpss.spatial; // Part of the spatial modelling package.
import java.awt.geom.*;   // For GeneralPath.

// ****************************************************************
/** Models a GIS vector object.
  * @author Jo Wood
  * @version 2.4, 21st October, 2001
  */
// ****************************************************************

public class GISVector extends SpatialObject
{
    // -------------------- Object variables --------------------

    private GeneralPath coords;    // Vector geometry array.
    private int type;              // Type of vector object.
    private float attribute;       // Attribute associated with vector.

    // --------------------- Constructor ----------------------
```

```java
    /** Creates an empty GIS vector object.
      */
    public GISVector()
    {
        type = POINT;
        attribute = 0;
        coords = new GeneralPath();
        setBounds(new Footprint(0,0,0,0));
    }

    /** Creates a GIS vector object with the given x and y coords.
      * @param xCoords Array holding x-coordinates of object.
      * @param yCoords Array holding y-coordinates of object.
      * @param type Type of vector object (POINT, LINE, AREA etc.)
      * @param attrib Attribute associated with this GIS vector.
      */
    public GISVector(float[] xCoords, float yCoords[],
                     int type, float attrib)
    {
        // Initialise vector information and check integrity.
        int numCoords = xCoords.length;
        this.type = type;
        this.attribute = attrib;
        coords = new GeneralPath();

        // Store the coordinates and calculate bounds of object.
        coords.moveTo(xCoords[0], yCoords[0]);

        for (int i=0; i<numCoords; i++)
        {
            coords.lineTo(xCoords[i],yCoords[i]);

            // Update bounds here.
        }

        if (type == AREA)
            coords.closePath();
    }
    // ----------------------- Methods -------------------------

    /** Adds the given footprint to the vector's coordinates and
      * updates the bounding area.
      * @param footprint Coordinates to add.
      */
    public void addCoords(Footprint fp)
    {
        coords.lineTo(fp.getXOrigin(), fp.getYOrigin());
        setBounds(getUnionMER(fp));
    }

    // -------------------- Accessor Methods --------------------

    /** Returns a drawable set of coordinates representing vector.
      * @return coordinates of the vector.
      */
    public GeneralPath getCoords()
    {
        return coords;
```

```
        }
        /** Reports the attribute associated with the GIS vector.
         *  @return Attribute associated with the GIS vector.
         */
        public float getAttribute()
        {
            return attribute;
        }

        /** Reports the attribute at the given point location.
         *  @param fp Location to query.
         *  @return Attribute associated with the GIS vector if given
         *  location intersects with object, otherwise OUT_OF_BOUNDS.
         */
        public float getAttribute(Footprint fp)
        {
            if (coords.contains(fp.getXOrigin(), fp.getYOrigin()))
                return attribute;
            else
                return OUT_OF_BOUNDS;
        }
    }
}
```

To add a coordinate pair to the collection stored in `coords`, the methods `moveTo()` and `lineTo()` are used. `moveTo()` adds a coordinate pair without 'joining' it to the previous point in the list. `lineTo()` draws a line from the previous point (if it exists) to the newly inserted one. This has the advantage of allowing us to build up more complex GIS vectors comprising multiple boundaries. It also has the considerable advantage of allowing us to employ some of Java2D's useful methods for handling 2D geometry.

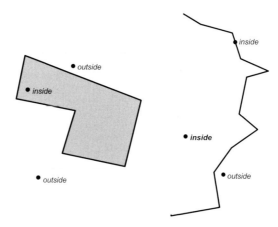

Figure 8.10 Point-in-polygon testing using `GeneralPath`'s `compare()` method. Note that the method returns a topologically incorrect result for concave linear features as line is transformed into an areal feature.

closePath() joins the first and last points in a coordinate collection to form a closed loop. This makes it appropriate for representing areal objects. contains() performs a 'point-in-polygon' search when given a coordinate pair to analyse. In other words, if we supply the method with a spatial footprint, it can tell us if that footprint is inside or outside the object represented by the GeneralPath. Unlike our own compare() method, this will not simply compare enclosing rectangles, but will perform a more accurate test around the outer boundary of the feature. We can use this method to return the attribute of any enclosing feature given a point-based footprint. By linking SpatialPanel's mouse monitoring, we now have a mechanism for reporting the attribute of any feature clicked with the mouse. The only problem with the contains() method used to do this, is that it will treat all objects as areas and so can give misleading results for linear features (see Figure 8.10).

8.3.4 The ModelDisplay GUI

We have now built almost all the necessary components to create a spatial model display and query GUI. We have classes for storing drawable GIS vectors and rasters, we have a set of panels for drawing each of the model types as well as monitoring mouse input. We also have a class (Header) for storing other textual information associated with a spatial object. Our final task is to assemble all these components inside a single JFrame.

We will create a JFrame with a window listener that allows us to close it cleanly, just as we did in Section 8.1.2.

```
package jwo.jpss.spatial.gui;    // Part of the spatial GUI package.

import jwo.jpss.spatial.*;        // For spatial classes.
import javax.swing.*;             // For Swing components
import java.awt.*;                // For AWT components.
import java.awt.event.*;          // For event handling.

// ****************************************************************
/** Creates a simple window for displaying spatial models.
  * @author    Jo Wood.
  * @version   1.3, 9th October, 2001
  */
// ****************************************************************

public class ModelDisplay extends JFrame
{
    // -------------------- Object Variables --------------------

    private JTextField  statusBar;
    private JTextArea   taInfo;
    private JPanel      p0;
    private SpatialObject spObject;

    // ---------------------- Constructor ----------------------
```

```java
    /** Creates a window which displays the given spatial object.
     * @param spObject Spatial object to display.
     */
    public ModelDisplay(SpatialObject spObject)
    {
        // Create closable window with a title and menu.
        super("Spatial Model Display");
        setDefaultCloseOperation(
                            WindowConstants.DO_NOTHING_ON_CLOSE);
        addWindowListener(new WinMonitor());
        Container contentPane = getContentPane();

        // Place everything inside a panel with spacing border.
        p0 = new JPanel(new GridLayout(1,2));
        p0.setBorder(BorderFactory.createEmptyBorder(4,4,4,4));
        p0.add(new JPanel());

        // Set up a text area to display model information.
        taInfo = new JTextArea();
        taInfo.setBackground(getBackground());
        taInfo.setFont(new Font("SansSerif", Font.BOLD, 12));
        taInfo.setEditable(false);
        taInfo.setLineWrap(true);
        taInfo.setWrapStyleWord(true);
        taInfo.setMargin(new Insets(4,8,4,4));
        JScrollPane scrollPane = new JScrollPane(taInfo);
        p0.add(scrollPane,1);

        // Create a status bar to report feedback.
        statusBar = new JTextField("Status:");
        statusBar.setBackground(getBackground());
        statusBar.setEditable(false);
        contentPane.add(statusBar,BorderLayout.SOUTH);

        // Add the model panel and a status bar.
        contentPane.add(p0, BorderLayout.CENTER);

        // Update panel with current spatial object.
        setSpatialObject(spObject);

        // Make the window visible.
        pack();
        setVisible(true);
    }


    // ---------------------- Methods ------------------------

    /** Updates the display with the given spatial object.
     * @param spObject New spatial object to display.
     */
    public void setSpatialObject(SpatialObject spObject)
    {
        this.spObject = spObject;
        SpatialPanel modelPanel = spObject.createPanel();
        modelPanel.setOutput(statusBar);

        Header header = spObject.getHeader();
        taInfo.setText(header.getTitle()+"\n\n");
```

```
        taInfo.append(spObject.getBounds()+"\n\n");
        taInfo.append("Author: "+header.getAuthor()+"\n\n");
        taInfo.append("Rights: "+header.getRights()+"\n\n");
        taInfo.append("Notes:  "+header.getNotes());

        p0.remove(0);
        p0.add(modelPanel,0);
    }


    // -------------------- Private Methods ---------------------

    /** Asks the user if they really want to quit, then closes.
      */
    private void closeDown()
    {
        int response = JOptionPane.showConfirmDialog(this,
                           "Are you sure you want to quit?");
        if (response == JOptionPane.YES_OPTION)
            System.exit(0);    // Exit program.
    }


    // -------------------- Nested Classes ----------------------

    /** Monitors window closing events and performs a 'clean exit'
      * when requested.
      */
    private class WinMonitor extends WindowAdapter
    {
        /** Responds to attempt to close window via the GUI. Checks
          * the user really wants to quite before closing down.
          * @param event Window closing event.
          */
        public void windowClosing(WindowEvent event)
        {
            closeDown();
        }
    }
}
```

Note that this class only makes use of methods found generally in `SpatialObject`, allowing it to display a `RasterMap`, `VectorMap` or indeed any spatial object we develop in the future. The header information associated with the object is displayed in a `JTextArea` rather than using `JLabels` as we did in Chapter 4. This gives us a little more control over the way in which the text is displayed (for example, by allowing text to wrap over multiple lines). A similar text area is placed along the south of the window, which will report any text information returned by the `SpatialPanel`. This provides valuable feedback when the mouse is moved or clicked over the spatial object. Examples of output are shown in Figure 8.11.

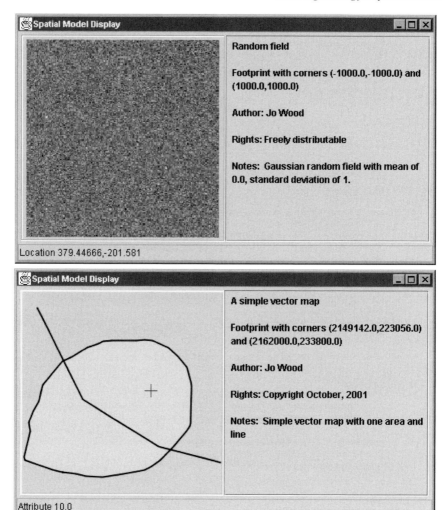

Figure 8.11 Examples of ModelDisplay output using a RasterMap and VectorMap.

8.4 SUMMARY

In this chapter, we have considered how we can incorporate a degree of dynamism in our classes. Event handling allows our classes to monitor changes at runtime and respond accordingly. A delegation model of event handling is used by the Swing and AWT graphical user interface classes where a component is made to listen out for events, and should one occur, it is given a reference to a class that will respond to that event.

We have also seen how we can model concurrent activities by creating multiple threads. Threads allow parallel processes to be modelled and are particularly useful when we wish to ensure a GUI remains responsive while part of a computer's resources are devoted to some other task.

We used the ideas of dynamic event handling to build a graphical user interface on top of the raster and vector spatial models. This involved using the Java2D package and some of its useful geometric handling functionality such as the affine transformation and high quality vector drawing.

By the end of this chapter and associated exercises on the web, you should be able to

- make buttons and menus respond to mouse selection;
- add other delegated event handling to your GUI classes;
- use nested adapter classes to simplify the process of handling events;
- create threaded processes by both inheriting the `Thread` class and implementing the `Runnable` interface;
- use the `Timer` class to schedule events at regular timed intervals; and
- understand how to use the Java2D classes to transform and draw high-quality vector graphics.

CHAPTER NINE

Handling Streams and Files

Way back in the second chapter, it was suggested that all programs consist of some input, process and output. This particular chapter will examine in more detail, some of the mechanisms for dealing with input – the gathering of information by the computer, and output – the reporting of information by the computer.

A computer can gather input in a variety of ways, some examples of which are given below:

- data typed in on the keyboard;
- data from a mouse/trackball/joystick;
- data from a microphone (e.g. voice recognition);
- data from a scanner/digitizer;
- data received from the internet; and
- data from a file on disk.

Likewise, the computer can report information to us (output) in a variety of ways:

- text displayed on screen;
- graphics displayed on screen;
- sound (music, sound effects, voice);
- send data over the internet;
- printer output; and
- create a file on disk.

In this chapter, we will consider how we can get our own Java programs to capture and write to some of these input and output devices.

9.1 INPUT AND OUTPUT STREAMS

Given that all computer processing involves dealing with binary numbers or *bits*, at some level, all of the examples above involve either sending or receiving long sequences of binary numbers. Such a sequence is known as a *stream*.

An input stream is a sequence of numbers coming into our program, while an output stream is a sequence of numbers sent out by our program. Things are made a little easier for us as Java programmers since all streams are dealt with in a similar fashion, regardless of the input or output device used. This chapter will consider some of the more common examples, particularly input from the keyboard, and input and output from and to files stored on disk. We will consider input and output over the internet in the following chapter.

9.1.1 Standard Keyboard Input

Several of the programs that we have written in these chapters have used the special class `KeyboardInput`. If you remember, this was a class written for this book that deals with prompting for, and converting text typed in using the keyboard, into numbers that may be used by our Java programs. It was used because it simplified the process of keyboard input when we wanted to concentrate on other issues.

Textual Input/Output with a Graphical User Interface

The discussion in this chapter concentrates on input and output without a GUI. For many programs, you can make use of the more friendly GUI classes for handling the input and output of text.

The most useful classes for text input in the Swing package are `JTextField` and `JTextArea`, both of which are inherited from the class JTextComponent. All contain the method `getText()` which returns a `String` containing the contents of whatever has been typed into these areas by the user.

As we have seen in several examples, information is often extracted from a text component when a button of some kind is pressed ('Submit', 'Calculate' 'OK' etc.). If more immediate response to text input is required, event handling can be used to monitor text input. Two of the more useful event listeners in this respect are the `TextListener` that can be added to any text component and `KeyListener`, which is used to monitor keyboard presses directly.

More sophisticated text-based output can be controlled using the components `JEditorPane`, `JTextPane` and `JFormattedTextField`. These components (also inherited from `JTextComponent`) allow formatting such as multiple fonts, emboldening, HTML and customised formatting instructions.

We are now ready to look at how input from the keyboard is actually processed by our Java programs. Although there are many ways in which information may be input into our Java programs, the default route, known as the *standard input stream*, is via the keyboard. In Java, this is referred to using `System.in`. There is a parallel here with screen output which we have been using since the first chapter, called `System.out` (followed by `println()` or `print()`). The standard keyboard input (and standard output) is 'standard' only in the sense of command line control of our programs (from the DOS prompt or Unix shell). As we have seen in previous chapters, we can create more user friendly input and output by

Handling Streams and Files

creating some form of graphical user interface with the AWT and Swing packages. However, it is still useful to be able to handle command line input and output for occasions when graphical interfaces are not necessary or possible.

To 'read' the values that have been typed in from the keyboard that arrive in our programs via `System.in`, we have to create an `InputStreamReader` object. This is created from a class that comes with the Java language contained in its own package `java.io`.

```
InputStreamReader inKeys = new InputStreamReader(System.in);
```

The `InputStreamReader` object requires us to send it a message telling it which stream to read from. In the example above, we tell it to read (standard) keyboard input, by specifying `System.in` as the parameter.

By default, a stream reader will read information from a stream one byte at a time. Given that we might want to send or receive many thousands of bytes of information, this can be a rather inefficient way of doing things. We can make the process of reading (and writing) streamed data much more efficient by *buffering* the data. In other words, the bytes of data in the stream are gathered together in larger groups before processing. This is achieved in Java by 'wrapping' the stream reader in a `BufferedReader`.

```
BufferedReader bufferedInKeys = new BufferedReader(inKeys);
```

These two lines are often combined in a single (and rather intimidating) line of code.

```
BufferedReader in = new BufferedReader(
                        new InputStreamReader(System.in));
```

The object `in` is now ready to receive anything typed in from the keyboard. To read that input, we call the method `readLine()` that is part of the `BufferedReader` class. This sends a `String` message back containing the text typed in from the keyboard.

```
String textLine = in.readLine();
```

We will see an example of how this can be used when we look at the `KeyboardInput` class below. But before we do that, we need to consider a new concept that is particularly important when handling input and output streams.

9.1.2 Catching Exceptions

You will have noticed that when you create Java programs, there are two types of errors that can occur.

The first and most common are *compile-time* errors. These errors occur because the Java compiler (javac) cannot convert your code into machine instructions (bytecode).

For example,

```
Sort2.java:38 Undefined variable: listChandged
     listChandged = true;
     ^
```

Contrary to what you may think, these types of error are the easiest to deal with since the Java compiler tells you what sort of error it has come across and which line in the program it lies.

The second class of errors occurs when your programs have been successfully compiled and you try to run them. These are known as *runtime* errors. They are more difficult to detect since it is usually less obvious what in your code has produced the error.

For example,

```
java.lang.NullPointerException
     at LandSerf.main(LandSerf.java:32)
```

Run-time errors, which in the Java language are known as *exceptions*, can often occur through no direct fault of the Java programmer. For example, if the computer runs out of memory, the hard disk fails to work, or the user types in an unexpected value. However, it is the programmer's responsibility to anticipate such errors by identifying parts of the code that may be associated with the generation of exceptions.

'Risky' parts of a program include, among others, creating large arrays (could run out of memory) and file input/output (could run out of disk space). Java allows the programmer to anticipate such risky parts of code using the keywords try and catch.

The general form is:

```
try
{
      // Some code that might generate an exception
}
catch (exceptionType exceptionName)
{
      // Some action to perform if exception produced
}
```

For example,

```
try
{
    textLine = in.readLine();
}
catch (IOException e)
{
    System.err.println("Problem reading input: " + e);
}
```

You will notice that the catch line 'catches' input/output errors known as IOExceptions. Despite its rather non-explanatory variable name, a common Java programming convention is to name all exceptions e. In this case, we just print out the exception preceded by the message Problem reading input:.

Note that you can place try{} around any section of code that is likely to generate exceptions, as long as there is at least one catch{} paired with it. In fact, Java will insist that you place a try/catch pair around any operation that uses file input/output.

This structure also allows the programmer to place an optional finally clause below the catch clause. Java will jump to this section whether or not an exception is generated.

```
String textLine;

try
{
    textLine = in.readLine();
}
catch (IOException e)
{
    System.err.println("Problem reading input: " + e);
}
finally
{
    System.out.println("Finished stream handling");
}
```

finally is generally used when we wish to perform some kind of 'clean up' operation after a section of risky code. This might involve, for example, closing opened streams or providing some status report on the success of the risky code.

One word of warning when using try and catch clauses. The clauses themselves are placed inside braces. Consequently, and variables or objects declared in these clauses will be treated as local variables, and cannot be used outside of the braces. It often therefore makes sense to *declare* variables (such as textLine above) outside of the clause, even if they are *initialised* within.

We can get a better idea of how stream input and exception handling is used by considering the `KeyboardInput` class that we have used in earlier programs.

```java
package jwo.jpss.utilities;
import java.io.*;              // For data input streams.

// ****************************************************************
/** Class to handle keyboard input. Prompts for user's keyboard
  * input and converts input string into numbers. Picks up
  * conversion errors.
  * @author    Jo Wood.
  * @version   1.4, 30th May, 2001
  */
// ****************************************************************

public class KeyboardInput
{
    // ----------------- Object variables ------------------

    private String textLine;   // Line of text input from keyboard.

    // ------------------- Constructor ---------------------

    /** Creates a keyboardInput object that handles prompting
      * and conversion of keyboard input.
      */
    public KeyboardInput()
    {
        textLine = new String();      // Create empty string.
    }

    // --------------------- Methods ---------------------

    /** Prompts the user to type in something from the keyboard.
      * Uses standard input to gether input.
      * @param promptText Message to prompt input with.
      */
    public void prompt(String promptText)
    {
        System.out.print(promptText + " ");
        textLine = readInput();
    }

    /** Extracts an integer number from the keyboard input.
      * @return Integer value extracted from keyboard input.
      */
    public int getInt()
    {
       int intVal=0;

       try
       {
          intVal = new Integer(textLine).intValue();
       }
       catch (NumberFormatException e)
       {
          System.err.println("Error converting input to integer.");
       }
```

Handling Streams and Files

```java
      return intVal;
   }

   /** Extracts a floating point number from the keyboard input.
    *  @return Floating point value extracted from keyboard input.
    */
   public float getFloat()
   {
      float floatVal=0.0f;

      try
      {
         floatVal = new Float(textLine).floatValue();
      }
      catch (NumberFormatException e)
      {
         System.err.println(
              "Error converting input to floating point number.");
      }
      return floatVal;
   }

   /** Extracts a double precision number from the keyboard input.
    *  @return Double precision value extracted from keyboard.
    */
   public double getDouble()
   {
      double doubleVal=0.0;

      try
      {
         doubleVal = new Double(textLine).doubleValue();
      }
      catch (NumberFormatException e)
      {
         System.err.println(
            "Error converting input to double precision number.");
      }
      return doubleVal;
   }

   /** Extracts a string typed in on the keyboard.
    *  @return String extracted from keyboard input.
    */
   public String getString()
   {
      return textLine;
   }

   /** Extracts the first letter typed in on the keyboard.
    *  @return First character extracted from keyboard input.
    */
   public char getChar()
   {
      char firstChar = ' ';

      try
      {
         firstChar = textLine.charAt(0);
```

```
            }
            catch (StringIndexOutOfBoundsException e)
            {
               System.err.println("Error finding first letter of input");
            }

            return firstChar;
      }

      // ------------------- Private Methods -----------------

      /** Returns a line of text input from standard input (usually
       * the keyboard).
       * @return Line of keyboard input.
       */
      private String readInput()
      {
            BufferedReader in = new BufferedReader(
                                    new InputStreamReader(System.in));
            try
            {
               textLine = in.readLine();
            }
            catch (IOException e)
            {
               System.err.println("Problem reading keyboard input.");
               return null;
            }
            return textLine;
      }
}
```

The KeyboardInput class uses several types of Exception handling. The first, which is used in the private method readInput handles input/output exceptions as discussed above. The class also anticipates two other forms of exception:

NumberFormatExceptions and StringIndexOutOfBoundsExceptions.

The first of these is generated if we try to convert a non-numerical string into a number. Since we cannot anticipate whether or not the user will type in text that is capable of being converted into a number, we need to handle such cases with an exception.

The second (StringIndexOutOfBoundsException) is used just in case the user presses the return key without typing any text when trying to extract the first character from their input.

9.2 READING AND WRITING FILES

So far, the Java programs we have created take their input either from the keyboard, or from variables defined within our programs. Likewise, output is sent

Handling Streams and Files 241

to the screen either graphically or as text. Quite frequently, we will require our programs to use more permanent sources of data input and output.

To write out, or read in, information to and from disk, we can get Java to manipulate *files*. In both cases, our programs must contain three separate steps.

1   Establish a link with a named file on disk
2   Read or write that file
3   Close the link with the file.

9.2.1 Opening a File

Opening a file for input is similar to the way we read keyboard input. For example, to open a file (called `input.txt`) from which we wish to read information:

`FileReader inFile = new FileReader("input.txt");`

To open a file (called `output.txt`) into which we wish to write information:

`FileWriter outFile = new FileWriter("output.txt");`

As with reading and writing to streams, we tend to encase such lines inside a buffered reader/writer and `try{} catch{}` clauses.

Opening a file for reading

```
BufferedReader inFile = null;

try
{
    inFile = new BufferedReader(new FileReader(fileName));
}
catch (FileNotFoundException e)
{
    System.err.println("Can't find file " + fileName + " :"+ e);
}
catch (IOException e)
{
    System.err.println("Error reading " + fileName + " :" + e);
}
```

Opening a file for writing

```
BufferedWriter outFile = null;

try
{
    outFile = new BufferedWriter(new FileWriter(fileName));
}
catch (FileNotFoundException e)
```

```
{
    System.err.println("Can't find file " + fileName + " :"+ e);
}
catch (IOException e)
{
    System.err.println("Error reading " + fileName + " :" + e);
}
```

9.2.2 Reading/Writing a File

Once a file has been opened, reading and writing from or to it is a relatively straightforward process. To read a line of text from a file, simply call the `readLine()` method that is part of `BufferedReader` class. For example,

```
String textLine = inFile.readLine();
```

To write a line of text to a file we use the `write()` method that is part of the `BufferedReader` class. For example,

```
String textLine = "message to write";
outFile.write(textLine,0,textLine.length());
```

The `write()` method takes three arguments as parameters. The first is the `String` containing the text to write to the file, the second is the position from the beginning of the text to start writing (nearly always 0), and the third is the length of the string which is conveniently found out by calling `String`'s `length()` method.

Note that both `readLine()` and `write()` keep track of the last point in the file that was touched. In other words, subsequent calls to the `write()` method will append text to the end of whatever was in the file. Likewise, subsequent calls of `readLine()` will read consecutive lines in a file. This raises the question of how do we know when we have come to the end of a file that we have been reading? The answer can be found by calling the method `ready()` in `BufferedReader`.

For example,

```
while (inFile.ready())
{
    textLine = inFile.readLine();
    System.out.println(textLine);
}
```

would read all the lines of text in a file and display them on screen. The program will iterate though the file, line by line while there are still some unread lines of text in the input file.

Handling Streams and Files 243

> **Text and Binary Files**
>
> What we have described in this section covers reading and writing **text** files. That is, files that are made up only of text characters (such as an HTML file). Java can also manipulate the individual bytes of a **binary** file. This is however a more complicated process and is generally not required by most Java programs.
>
> We will however see a little later on, how we can read and write entire objects as binary files from and to disk using the process of **object serialization**

9.2.3 Closing a File

The final (and easiest) stage of file handling is to close the file we have been working with. We do this by calling the `close()` method of our `BufferedReader` or `BufferedWriter` object.

For example,

```
outFile.close();
inFile.close();
```

It is important to close files after using them. If you forget, your program may appear to work as expected, but the unclosed files will sit on the computer occupying valuable resources. In some cases, files that have not been closed will not be accessible to other applications, or even appear completely empty.

9.2.4 Opening Files with a Graphical User Interface

Creating a Graphical User Interface (GUI) does not obviate the need to read and write files, but it does greatly simplify the process of getting the user to choose the name of a file name to read or write. The Swing package contains a class `JFileChooser` designed to provide a platform-neutral GUI for selecting files within a file system.

Once an object is created from the `JFileChooser` class, the methods `showOpenDialog()` and `showSaveDialog()` can be called to display a graphical file chooser on the screen. The example below shows how this class can be used to select a file for reading. Note that the file chooser itself does not open or read any files, it simply allows a file name to be chosen by the user. Figures 9.1 and 9.2 show examples of output from this program.

```
import javax.swing.*;        // For file chooser GUI.
import java.io.*;            // For file handling.
//   ***************************************************************
```

```java
/** Asks the user for a file to load. Demonstrates the
  * use of the GUI component FileChooser to select files.
  * @author Jo Wood.
  * @version 1.1, 27th September, 2001.
  */
// ***************************************************************

public class FileBrowse extends JFrame
{
    // ------------------ Object Variables --------------------

    private JFileChooser fileChooser;

    // -------------------- Constructor ----------------------

    /** Asks user to select a file to read. Checks that the file
      * really exists.
      */
    public FileBrowse()
    {
        // Initialise application window and file chooser.
        super();
        fileChooser = new JFileChooser();
        BufferedReader bInFile = null;

        // Ask user for a file to open.
        int choice = fileChooser.showOpenDialog(this);
        if (choice != JFileChooser.APPROVE_OPTION)
            return;

        // Retrieve filename and attempt to open it.
        File inFile = fileChooser.getSelectedFile();

        try
        {
            bInFile = new BufferedReader(new FileReader(inFile));
        }
        catch (FileNotFoundException e)
        {
            JOptionPane.showMessageDialog(this,
                        inFile.getAbsolutePath()+" not found.",
                        "Problem Opening File",
                        JOptionPane.ERROR_MESSAGE);
            return;
        }
        catch (IOException e)
        {
             JOptionPane.showMessageDialog(this,
                        "Error reading "+inFile.getAbsolutePath(),
                        "Problem Reading File",
                        JOptionPane.ERROR_MESSAGE);
            return;
        }

        // File reading would go here.

        // Close file.
        try
        {
```

```
            bInFile.close();
    }
    catch (IOException e)
    {
            JOptionPane.showMessageDialog(this,
                       "Error closing "+inFile.getAbsolutePath(),
                       "Problem Closing File",
                       JOptionPane.ERROR_MESSAGE);
    }
  }
}
```

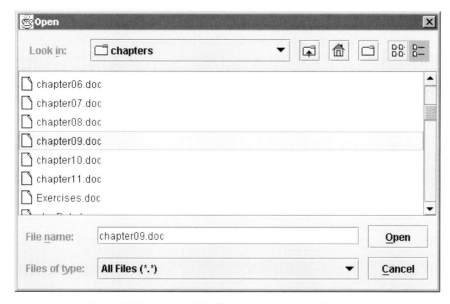

Figure 9.1 Example of a File Chooser used to select a file to open.

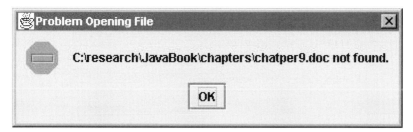

Figure 9.2 Example of error message reported if selected file is not found.

The file chooser contains both an 'Open' and 'Cancel' button allowing the user to back out of file selection if they wish. It is therefore necessary to test which button was selected by the user before proceeding with any file manipulation. Assuming the 'Cancel' button was not pressed, the chosen file can be retrieved using the `getSelectedFile()` method. This returns an object of type `File` that is used for storing the full name and path of any arbitrary file. `File` objects may be passed directly to a `FileReader` / `FileWriter`, or the name may be extracted using `getAbsolutePath()`.

If a file chooser is to be used more than once in a program (as is usually the case), it is good practice to make the `JFileChooser` an object variable, initialised only once. That way, whenever the chooser is displayed by the user, it will 'remember' the last directory chosen. This is the type of 'obvious' behaviour expected by most users of GUIs and helps to make file choosing a more intuitive process.

Occasionally, it is necessary to limit the choice of files available for either reading or writing. This may be achieved graphically by creating a *file filter* that only displays a restricted set of files inside a chooser. To create a file filter, it is necessary to extend Swing's `FileFilter` class and override the method that determines if any given file can be displayed. The example below shows how a filter to display only .gif and .jpg (image) files can be created.

```java
import javax.swing.filechooser.*;
import java.io.File;

//    ****************************************************************
/** Provides a file filter to allow only images to be displayed.
 *  @author   Jo Wood.
 *  @version 1.1, 30th September, 2001.
 */
//    ****************************************************************

public class ImageFileFilter extends FileFilter
{
    // ------------------- Implemented Methods ------------------

    /** Identifies jpg and gif files as visible to a file chooser.
     *  @param file File to consider displaying.
     *  @return true if file is an image file in current directory.
     */
    public boolean accept(File file)
    {
        if (file.isDirectory())
            return true;

        // Find file name extension.
        String fileName = file.getName();
        int i = fileName.lastIndexOf('.');

        if ((i > 0) && (i < fileName.length()-1))
            if ((fileName.substring(i+1).equalsIgnoreCase("jpg")) ||
                (fileName.substring(i+1).equalsIgnoreCase("gif")))
                return true;

        return false;
    }
```

Handling Streams and Files 247

```
    /** Reports description of this filter for the file chooser.
     * @return Description of this filter.
     */
    public String getDescription()
    {
        return new String("Image files");
    }
}
```

This filter can be added to the `JFileChooser` using the method `setFileFilter()`. For example,

`fileChooser.setFileFilter(new ImageFileFilter());`

Output from a file chooser using this filter is shown in Figure 9.3.

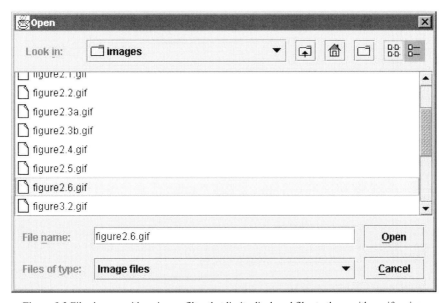

Figure 9.3 File chooser with an image filter that limits displayed files to those with a .gif or .jpg extension.

9.3 ANALYSING STRINGS

The procedure described in Section 9.2 for reading in files does so by reading text from a file line by line. Suppose our file contained the following information:

```
This is the first line of a text file.
This is the second line.
And a final closing line.
```

If our program simply treated each line of text as a single string of characters, the methods described above would be sufficient. However, quite commonly, we may wish to analyse the individual words within each line. To do this we need to create what is called a *String Tokenizer*.

A string token is simply a part of a string we are interested in. Usually, this will be a word within the string. To define a token, we also need to define the type of characters that separate them (known as *delimiters*). By default, the delimiters are spaces, tabs, and new lines.

To use a string tokenizer in Java consider the following code fragment.

```
import java.util.*;    // Required to use the string tokenizer.

public void wordCount()
{
    String textLine = "This is the first line of a text file";
    StringTokenizer sToken = new StringTokenizer(textLine);
    int word = 1;

    while (sToken.hasMoreTokens())
    {
        System.out.println(word+". "+sToken.nextToken());
        word++;
    }
}
```

The StringTokenizer object is created from the class by supplying the string to analyse as a parameter. The two important methods that are used once the StringTokenizer object has been created are hasMoreTokens() and nextToken(). These are used to iterate through the string, token by token.

hasMoreTokens() will return either true or false depending on whether there is any text left in the string to analyse. nextToken() will start from the current search position in the string (initially at the beginning) and return a String message containing the next token. Calling this method results in the search position moving to the end of the token just extracted.

The result of running the code shown above is to print out each word in textLine, one by one:

1. This
2. is
3. the
4. first
5. line
6. of
7. a
8. text
9. file.

Handling Streams and Files

You can use `StringTokenizers` to analyse the contents of text stored in files, or from keyboard input. It provides a very powerful way of parsing text messages, and complex text files. The case study at the end of this chapter demonstrates how string tokenizers can be used to import text files representing GIS vectors and rasters.

9.4 OBJECT SERIALIZATION

So far, all of the files we have dealt with have been text files. That is, they store strings of characters. This may be suitable for many applications, but they do have their limitations.

Suppose we wish to store as a file, an object of class `RasterMap` created in Chapter 7. Recall that this class contains information such as the title, dimensions, location and raster values. It would be possible to convert each of these items of information into `Strings` and save them in a file, but this would be a rather inefficient and cumbersome process. If we wish to read in such a file in order to create a new raster, we would have to extract the relevant text from the file in order to create our new `Raster` object.

A much simpler solution is to save the `RasterMap` object directly to a file. That is, take the section of computer's memory storing the raster object and write it out to disk. Likewise, it is more convenient to read the file from disk, and place it directly into memory instantly 'reconstructing' the object ready for use by our Java program.

This process can be relatively easily achieved in Java, and is known as *object serialization*. To allow an object to be serialized, we need to implement a new interface in its class definition illustrated in the example below.

```
import java.io.*;   // Package containing Serializable interface.

public class RasterMap implements Serializable
{
    // Class definition as usual.
}
```

Unusually for an object-oriented interface, `Serializable` does not contain any field or method definitions. It simply serves as an identifier indicating to Java that we wish to allow the class that implements the interface to be able to be serialized.

Once a class has been made serializable, we can simply save or read an object of that class straight to or from a file. The following method provides an example.

```
public boolean writeObject()
{
    try
    {
```

```java
        // Serialize the raster map.

        FileOutputStream oStrm = new FileOutputStream(fileName);
        BufferedOutputStream oBStrm =new BufferedOutputStream(oStrm);
        ObjectOutput oOStrm = new ObjectOutputStream(oBStrm);

        oOStrm.writeObject(rasterMap);
        oOStrm.flush();
        oOStrm.close();
    }
    catch (FileNotFoundException e)
    {
        System.err.println("Can't find " + fileName + " :"+ e);
        return false;
    }
    catch (IOException e)
    {
        System.err.println("Error writing " + fileName + " :" + e);
        return false;
    }
    return true;
}
```

This example assumes we have object variables `fileName` and `rasterMap`, which are the `String` file name and `RasterMap` objects respectively.

To read a serialized file back into an object, we would create a similar method.

```java
public boolean readObject()
{
    try
    {
        FileInputStream iStrm = new FileInputStream(fileName);
        BufferedInputStream iBStrm = new BufferedInputStream(iStrm);
        ObjectInput ioStream = new ObjectInputStream(iBStrm);

        rasterMap = (RasterMap)(ioStream.readObject());
        ioStream.close();
    }
    catch (IOException e)
    {
        System.err.println("Error reading " + fileName + " :" + e);
        return false;
    }
    catch (ClassNotFoundException e)
    {
        System.err.println("Unknown format " + fileName + " :" + e);
        return false;
    }
    catch (ClassCastException e)
    {
        System.err.println("Unknown format " + fileName + " :" + e);
        return false;
    }
    return true;
}
```

Handling Streams and Files 251

When a serialised object is 'reconstructed', we need to inform Java the class into which it should attempt to fit the object. This is an example of typecasting that we have used many times before. However, in this case, the programmer does not have any control over the file that the user of the program might select. It therefore makes sense to place the deserialization in a `try-catch` clause and test for `ClassCastExceptions`. This code may seem a little intimidating at first, but the serialization of objects provides a very convenient way of storing Java objects as files on disk, or even as streams sent over the internet.

9.5 CASE STUDY: ADDING FILE HANDLING TO SPATIAL OBJECTS

This case study focuses on adding some of the file input and output functionality discussed in this chapter to the spatial models we have been developing. In particular, we will do the following

- Allow `VectorMaps` and `RasterMaps` to be saved to and loaded as files;
- provide an import facility to convert text files into raster and vector maps; and
- develop the spatial model GUI to allow file input/output via menu selections.

9.5.1 Importing Text Files

The `RasterMap`, `VectorMap` and `SpatialObject` classes we have developed so far can only receive input via their constructors. This has limited the volume of data they can store to whatever can be generated within our Java programs. Now that we have considered how to read lines of text from a file stored on disk, we can add methods to `RasterMap` and `VectorMap` for importing spatial data from text files.

File Formats

Spatial data are stored in many hundreds of different file formats, often making *interoperability* between software packages difficult. We will consider two simple file formats, one for storing raster data, the other vector data.

The *ASCII ArcGrid* format provides a simple way of representing a raster map as a text file. This format is used by the GIS *ArcInfo* when exporting raster data in text format and is described in Table 9.1.

The *Generic VectorMap* format, described in Table 9.2, is used by the GIS *GRASS* and can be used to represent both the geometry and simple numerical attributes of any 2D vector objects.

Table 9.1 The *ArcGrid* ASCII file format

Example	Description

`ncols 321`
`nrows 468`
`xllcorner 387570.0`
`yllcorner 5289240.0`
`cellsize 30`
`nodata_value -32766`
`-32766 393 393 393 393 ..etc`
`-32766 395 396 396 394 ..etc`
`:`
`etc.`
	Header information consists of at least the number of rows/columns, the geographical origin and the grid cell resolution. The code representing missing values can also be defined (`nodata_value`).
	
	Raster data are stored in *nrows* rows from north to south, each row consisting of *ncols* values from west to east.
	
	Raster values can be floating point or integer numbers and are separated by any number of whitespace characters.

Table 9.2 Generic (*GRASS*) vector map file format

Example	Description

`P 340000.0 483520.0 5`
`P 345000.0 479050.0 10`
`L 2 7`
`366000.0 489530.0`
`375210.0 488100.0`
`A 3 9`
`366000.0 489550.0`
`368100.0 488780.0`
`368200.0 488180.0`
`:`
`etc.`
	No header information is used. Feature type is identified by a P, L or A prefix representing point, line and area features respectively.
	
	Point data are stored in `x y Attrib` order, one point per line.
	
	Line and Area data have an initial line containing the number of coordinates and a feature attribute. Coordinates follow in x y order.
	
	All coordinates and attributes can be in either integer or floating point formats.

Although we will add methods to import both of these formats into our spatial classes, we can anticipate importing further formats in the future, so it makes sense to create a new separate class that handles import translation. We will keep this new class as independent as possible from the other spatial classes by making the methods within it static. Classes that consist of entirely static methods like this are sometimes called *helper classes*. The structure of the file handling helper class is shown below.

```
package jwo.jpss.spatial; // Part of the spatial modelling package.
import java.io.*;         // For file handling.

// ***************************************************************
/** Provides a series of static methods for handling GIS files.
  * @author Jo Wood.
  * @version 2.1, 9th October, 2001
  */
// ***************************************************************
```

```java
public class FileHandler
{
    // -------------------- Class variables ---------------------
                            /** Unknown file format.*/
    public static final int UNKNOWN = 1;
                            /** Internal file format.*/
    public static final int INTERNAL = 1;
                            /** ArcGrid ASCII raster format.*/
    public static final int ARC_GRID_ASCII = 2;
                            /** Generic ASCII vector format.*/
    public static final int VECTOR_ASCII = 3;

    // ---------------------- Methods --------------------------

    /** Attempts to read the given spatial file.
      * @param inFile File to read.
      * @return Read spatial object or null if file not read.
      */
    public static SpatialObject readFile(File inFile)
    {
        int fileType = guessFileType(inFile.getName());
        return readFile(inFile,fileType);
    }

    /** Attempts to read the given spatial file of the given type.
      * @param inFile File to read.
      * @param fileType Type of file to read.
      * @return Read spatial object or null if not read.
      */
    public static SpatialObject readFile(File inFile, int fileType)
    {
        switch (fileType)
        {
            case(INTERNAL):
                return readInternal(inFile);
            case(ARC_GRID_ASCII):
                return readArcGrid(inFile);
            case(VECTOR_ASCII):
                return readVectText(inFile);
            default:
                return null;
        }
    }

    /** Attempts to predict type of file based on its extension.
      * @param fileName Name of file to guess.
      * @return Type of file, or UNKNOWN if not known.
      */
    public static int guessFileType(String fileName)
    {
        // Find file filename extension.
        int i = fileName.lastIndexOf('.');

        if ((i <= 0) || (i > fileName.length()-1))
            return UNKNOWN;      // No filename extension.

        String ext = fileName.substring(i+1);
```

```java
            if (ext.equalsIgnoreCase("gis"))
                return INTERNAL;
            if (ext.equalsIgnoreCase("grd"))
                return ARC_GRID_ASCII;
            if (ext.equalsIgnoreCase("txt"))
                return VECTOR_ASCII;

            return UNKNOWN;
    }

    // ------------------- Private Methods --------------------

    /** Attempts to read given file representing an ArcGrid raster.
      * @param inFile File to read.
      * @return Read spatial object or null if not read.
      */
    private static SpatialObject readArcGrid(File inFile)
    {
        // File reading code here.
    }

    /** Attempts to read given file representing ascii vector map.
      * @param inFile File to read.
      * @return Read spatial object or null if not read.
      */
    private static SpatialObject readVectText(File inFile)
    {
        // File reading code here.
    }
}
```

The class provides two `readFile()` methods, requiring at least the name of the file to read, and optionally, an explicit reference to the file format. If a file format is not given, the method `guessFileType()` is used to guess the format based on the filename extension. Currently, this class assumes that ArcGrid files will have the .grd extension and generic vector files the .txt extension.

Importing a Generic Vector Text File

Since the format used for importing vector maps is entirely text based, we can use a `FileReader` and `StringTokenizer` to process its contents.

```java
/** Attempts to read given file representing ascii vector map.
  * @param inFile File to read.
  * @return Read spatial object or null if not read.
  */
private static SpatialObject readVectText(File inFile)
{
    VectorMap vectorMap=new VectorMap();
    try
    {
        BufferedReader bInFile = new BufferedReader(
                                       new FileReader(inFile));
        StringTokenizer sToken;    // Reads in word at a time

        while (bInFile.ready())
```

```java
        {
            int numCoords = 1;
            int type = SpatialModel.POINT;
            float x[],y[],attr;
            sToken = new StringTokenizer(bInFile.readLine());
            String word = sToken.nextToken().toUpperCase();

            if (word.startsWith("P"))
            {
                x = new float[1];
                y = new float[1];
                x[0]=Float.valueOf(sToken.nextToken()).floatValue();
                y[0]=Float.valueOf(sToken.nextToken()).floatValue();
                attr=Float.valueOf(sToken.nextToken()).floatValue();
                vectorMap.add(new GISVector(x,y,type,attr));
                continue;
            }
            else if (word.startsWith("L"))
            {
                numCoords = Integer.parseInt(sToken.nextToken());
                type = SpatialModel.LINE;
            }
            else if (word.startsWith("A"))
            {
                numCoords = Integer.parseInt(sToken.nextToken());
                type = SpatialModel.AREA;
            }
            else continue;

            attr = Float.valueOf(sToken.nextToken()).floatValue();
            x = new float[numCoords];
            y = new float[numCoords];
            for (int i=0; i<numCoords; i++)
            {
                sToken = new StringTokenizer(bInFile.readLine());
                x[i]=Float.valueOf(sToken.nextToken()).floatValue();
                y[i]=Float.valueOf(sToken.nextToken()).floatValue();
            }
            vectorMap.add(new GISVector(x,y,type,attr));
        }
    }
    catch (FileNotFoundException e) {return null;}
    catch (IOException e) {return null;}
    catch (NumberFormatException e) {return null;}

    return vectorMap;
}
```

The method reads the file line by line and constructs a string tokenizer to evaluate each 'word' in each line. Each new GIS vector should be represented by a line starting with P, L or A, depending on its dimensions. The method therefore searches for one of these letters at the start of a line. If it finds one, it proceeds to read in the coordinates, attributes and possibly the number of coordinates in the object (for line and areas only). The Java keyword continue is used to skip any unexpected lines.

After the coordinates of a GIS vector have been read, a `GISVector` object is created from them and added to the `VectorMap`. This process continues until there are no more lines in the file to read. Finally, the imported `VectorMap` is returned as an outgoing message.

A similar method can be created for reading ArcGrid text files, again examining the file line by line, extracting the tokens of each line, and populating a `SpatialModel`, which in this case would be a `RasterMap`.

9.5.2 Making Spatial Objects Serializable

We now have a facility for importing externally produced spatial objects, but as yet no way of saving a copy of a spatial object to disk. We could add some methods to the `FileHandler` class that exported a given spatial model as a text file, but this would be rather inefficient both in terms of file space required to store text and in having to write separate methods for saving raster and vector models. The more efficient and object-oriented approach would be to save the spatial object in its serialized form. This means that raster, vector and any future spatial objects can all be handled in an identical way.

As we considered in Section 9.4, making an object serializable is quite straightforward – the serializable class simply implements the `java.io.Serializable` interface. When we make a class serializable in this way, any of its subclasses will automatically become serializable too. So, to make both `RasterMap` and `VectorMap` capable of being serialized, we get `SpatialObject` (their superclass) to implement `Serializable`.

There are however, restrictions on what can be serialized. If a class is to be serializable, so must all of its object variables. This means that both `Footprint` and `Header` must also implement `Serializable` (they are both object variables inside `SpatialObject`).

Figure 9.4 shows the revised class diagram of our spatial classes incorporating the new serializable interface and file handling classes.

Now that all the classes we wish to save can be serialized, we can add an open and save method to the `FileHandler` class.

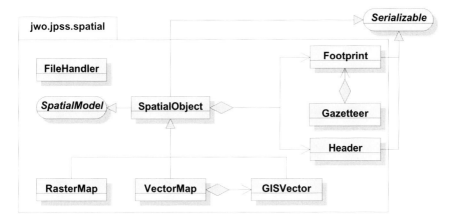

Figure 9.4 Spatial class diagram with serialization.

```
/** Attempts to save the given spatial object as the given file.
  * @param spObject Spatial object to save.
  * @param outFile File name to save.
  * @return true if saved correctly.
  */
public static boolean saveFile(SpatialObject spObject,
                               File outFile)
{
    try
    {
        FileOutputStream oStrm = new FileOutputStream(outFile);
        BufferedOutputStream oBStrm=new BufferedOutputStream(oStrm);
        ObjectOutput oOStrm = new ObjectOutputStream(oBStrm);

        oOStrm.writeObject(spObject);
        oOStrm.flush();
        oOStrm.close();
    }
    catch (FileNotFoundException e) { return false; }
    catch (IOException e) { return false; }

    return true;
}
```

The method for reading serialized files is private as this is only called by the public method `readFile()` if the filename has a `.gis` extension. Should we wish to add other formats for reading files in the future, we might choose to make the method public in a similar way to the `writeFile()` method.

```
/** Attempts to read given file representing a serialized object.
  * @param inFile File to read.
  * @return SpatialObject containing read file or null if not read.
  */
private static SpatialObject readInternal(File inFile)
{
```

```
    SpatialObject spObject;

    try
    {
        FileInputStream iStrm = new FileInputStream(inFile);
        BufferedInputStream iBStrm= new BufferedInputStream(iStrm);
        ObjectInput ioStream = new ObjectInputStream(iBStrm);

        spObject = (SpatialObject)(ioStream.readObject());
        ioStream.close();
    }
    catch (IOException e) {return null;}
    catch (ClassNotFoundException e) {return null;}
    catch (ClassCastException e) {return null;}

    return spObject;
}
```

Controlling Serialization

We now have a mechanism for importing raster and vector text files, and opening and saving spatial objects in serialized form. However, we have one remaining problem with the GISVector class. This class needs to be serialized as it is used as an object variable inside VectorMap (see Figure 9.4). Since GISVector inherits SpatialObject, it should be automatically serialized. Yet GISVector itself contains an object variable of class GeneralPath (see Section 8.3.3). GeneralPath, which forms part of the Java2D toolkit is not serializable. Furthermore, it is a final class, meaning we cannot subclass it into a serialized version. If we run the program as it stands and attempt to save a VectorMap, Java would issue the following runtime exception

java.io.NotSerializableException: java.awt.geom.GeneralPath

What we require is more control over what is serialized and what is not within the GISVector class. Thankfully, Java provides a mechanism to prevent certain classes from being serialized with the modifier transient. Any object variable declared with this keyword will be excluded from the serialization process. This eliminates the runtime error above, but prevents the vector geometry from being saved or loaded. To overcome this problem, we need to create a serializable variable to store geometry (for example, an array of Footprints). Yet storing geometry twice, once as a GeneralPath and once as an array of Footprints is wasteful of resources. What we require is a procedure that will transfer the geometry to and from a GeneralPath *only at the point of serialization and deserialization.*

This can be achieved by overriding the methods readObject() and writeObject() and customising what is and is not serialized. This is a useful process when we wish to control exactly what is written to and read from a file. It can be used for example to encrypt sensitive data or to compress/decompress large data structures. In our case, we will use it to transfer geometry between the unserializable GeneralPath and the serializable array of Footprints.

Customised serialization methods in `GISVector`

```
private transient GeneralPath coords;      // Vector geometry array.
private Footprint serializedCoords[];

/** Serializes the GISVector. Converts the general path into a
  * coordinate array.
  * @param s Stream to place serialization.
  */
private void writeObject(ObjectOutputStream s) throws IOException
{
    // Save general path as an array of footprints.
    PathIterator i = coords.getPathIterator(new AffineTransform());
    float segment[] = new float[6];
    serializedCoords = new Footprint[numCoords];

    for (int seg=0; seg<numCoords; seg++)
    {
        i.currentSegment(segment);
        serializedCoords[seg]=new Footprint(segment[0],segment[1]);
        i.next();
    }

    s.defaultWriteObject();  // Carry out serialization as normal.
    serializedCoords = null;
}


/** Deserializes the GISVector. Converts the coordinate array into
  * a general path.
  * @param s Stream from which to read serialization.
  */
private void readObject(ObjectInputStream s)
                       throws IOException, ClassNotFoundException
{
    s.defaultReadObject();  // Carry out deserialization as normal.

    // Reconstruct general path from coordinate array.
    coords = new GeneralPath();
    if (numCoords == 0)
        return;

    coords.moveTo(serializedCoords[0].getXOrigin(),
                  serializedCoords[0].getYOrigin());

    for (int i=0; i<numCoords; i++)
    {
        coords.lineTo(serializedCoords[i].getXOrigin(),
                      serializedCoords[i].getYOrigin());
    }

    if (type == AREA)
        coords.closePath();

    serializedCoords = null;
}
```

serializedCoords is the object variable storing the array of Footprints. In order to save on system resources, this array is removed (by setting it to null) after the conversion process has taken place, so that for most of GISVector's life, it stores only one copy of its geometry. To extract the geometry from a GeneralPath, we need to create a PathIterator which allows us to extract the sequence of (x, y) pairs that make up the vector boundary.

9.5.3 Developing the Graphical User Interface

The final task in adding input/output functionality to our spatial classes, is to provide a GUI to the file handling capabilities. We will add menu options to ModelDisplay allowing files to be opened and saved.

File handling in the ModelDisplay class

```
package jwo.jpss.spatial.gui;   // Part of the spatial GUI package.
//   ****************************************************************
/** Creates a simple window for displaying spatial models.
  * @author    Jo Wood.
  * @version   1.3, 9th October, 2001
  */
//   ****************************************************************
public class ModelDisplay extends JFrame implements ActionListener
{
    // -------------------- Object Variables --------------------

    private JMenuItem mOpen,mSave,mExit;
    private JFileChooser fileChooser;

    // ---------------------- Constructor ----------------------

    /** Creates a window which displays the given spatial object.
      * @param spObject Spatial object to display.
      */
    public ModelDisplay(SpatialObject spObject)
    {
        JMenuBar menuBar = new JMenuBar();
         JMenu menFile = new JMenu("File");
          mOpen = new JMenuItem("Open...");
          mOpen.addActionListener(this);
         menFile.add(mOpen);
          mSave = new JMenuItem("Save...");
          mSave.addActionListener(this);
         menFile.add(mSave);
         menFile.addSeparator();
          mExit = new JMenuItem("Exit");
          mExit.addActionListener(this);
         menFile.add(mExit);
        menuBar.add(menFile);
        setJMenuBar(menuBar);
```

```java
        // Initialse the file chooser.
        fileChooser = new JFileChooser();
        fileChooser.setFileFilter(new SpatialFileFilter());

        // Initialisation code here.
    }

    // ---------------------- Methods -------------------------

    /** Attempts to open given file and update the spatial object
      * being displayed.
      * @param inFile Spatial file to open.
      * @return True if opened successfully.
      */
    public boolean openFile(File inFile)
    {
        SpatialObject spObject = FileHandler.readFile(inFile);
        if (spObject==null)
        {
            statusBar.setText("Problem opening " +
                                        inFile.getAbsolutePath());
            return false;
        }
        setSpatialObject(spObject);
        statusBar.setText("New spatial object: " +
                                        inFile.getAbsolutePath());
        return true;
    }

    /** Attempts to save the current spatial object.
      * @param outFile Spatial file to save.
      * @return True if saved successfully.
      */
    public boolean saveFile(File outFile)
    {
        if (FileHandler.saveFile(spObject, outFile) == false)
        {
            statusBar.setText("Problem saving " +
                                        outFile.getAbsolutePath());
            return false;
        }
        statusBar.setText("Saved: " + outFile.getAbsolutePath());
        return true;
    }

    // ----------------- Implemented Methods --------------------

    /** Responds to a selection of menu item.
      * @param event Menu selection event.
      */
    public void actionPerformed(ActionEvent event)
    {
        if (event.getSource() == mExit)
            closeDown();

        if (event.getSource() == mOpen)
        {
            int choice = fileChooser.showOpenDialog(this);
```

```
                    if (choice == JFileChooser.APPROVE_OPTION)
                        openFile(fileChooser.getSelectedFile());
            }

            if (event.getSource() == mSave)
            {
                int choice = fileChooser.showSaveDialog(this);
                if (choice == JFileChooser.APPROVE_OPTION)
                    saveFile(fileChooser.getSelectedFile());
            }
        }
    }
```

Menu selection events are listened for and when necessary, the relevant file opening or saving method is called. A `JFileChooser` is used to allow the user in select files graphically. In this case, the file chooser is initialised with a *file filter* allowing only certain files to be displayed. This customised filter, created by extending the Swing class `FileFilter`, limits displayed files to those with a .gis, .grd or .txt extension.

```
package jwo.jpss.spatial.gui;

import javax.swing.filechooser.*;
import java.io.File;

// ****************************************************************
/** Provides a file filter to allow only spatial files to be
  * to be displayed in a JFileChooser.
  * @author  Jo Wood.
  * @version 1.2, 30th September, 2001.
  */
// ****************************************************************

public class SpatialFileFilter extends FileFilter
{
    // --------------------- Class variables ------------------

    private static final String[] extensions = {"gis","grd","txt"};

    // ------------------ Implemented Methods ----------------

    /** Filters only file extensions of known spatial file formats.
      * @param file File to consider displaying.
      * @return true if file has a recognised filename extension.
      */
    public boolean accept(File file)
    {
        if (file.isDirectory())
            return true;

        // Find file name extension.
        String fileName = file.getName();
        int i = fileName.lastIndexOf('.');

        if ((i > 0) && (i < fileName.length()-1))
        {
```

```
                String ext = fileName.substring(i+1);
                for (int extNum=0; extNum<extensions.length; extNum++)
                {
                    if (ext.equalsIgnoreCase(extensions[extNum]))
                        return true;
                }
            }
            return false;
        }

        /** Reports description of this filter for the file chooser.
          * @return Description of this filter.
          */
        public String getDescription()
        {
            return new String("Spatial files");
        }
    }
```

We now have fully functional spatial object display class capable of importing, displaying and saving spatial models in a range of formats. Sample output is shown in Figure 9.5.

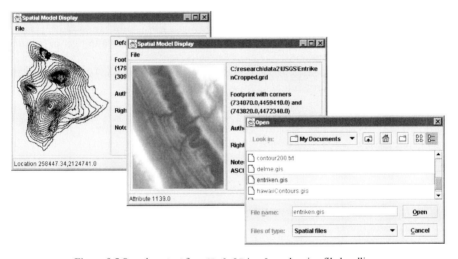

Figure 9.5 Sample output from `ModelDisplay` showing file handling.

9.6 SUMMARY

In this chapter, we have seen how to handle keyboard input directly in our Java programs. This can be useful when we do not wish to construct a graphical user interface, but still require input from the user of our programs. Handling keyboard input is similar to the way in which files are read from a file system. File handling as well as input and output from and to a range of devices can be handled by creating streams along which data are routed.

We have also seen that any code that runs the risk of generating runtime exceptions, such as out of memory or disk writing errors, can be caught and dealt with using the `try{}` and `catch{}` clauses.

Analysing text input, whether from a file or directly from the keyboard is made much easier by breaking down strings into words. The `StringTokenizer` class provides a convenient mechanism for doing this.

Finally, we saw that one of the easiest and most efficient ways of storing information on disk is to use *object serialization* for reading and writing objects. This allows us to transfer objects directly between some input or output device and the computer's memory. This was illustrated by adding file opening and saving functionality to the spatial model display classes developed previously.

By the end of this chapter and associated exercises on the web, you should be able to

- create a Java class that reads text input from the keyboard;
- open, read, write and close files for storing textual information;
- analyse the contents of a text string;
- write methods for reading and writing text files in a range of simple formats; and
- serialize and deserialize any object by implementing a `Serializable` interface and creating an appropriate object stream.

CHAPTER TEN

Communicating with the Wider World

One of the reasons Java has received so much popular attention as a programming language is the ease with which it can be used over the internet. From the outset of its design, it was intended that Java could be embedded in web pages allowing communication between server and client platforms. The ease with which this can be done was helped by its platform independence.

Over time, new communication technologies have been added to the core set of Java classes. These allow communication with relational databases, with remote Java virtual machines, with XML and other technologies. This chapter considers some of these communication mechanisms and shows how Java may be used to facilitate such communication.

10.1 THE APPLET

One of the simplest (and widely used) mechanisms for sharing Java programs over the internet is through the creation of Java applets. In general terms, an applet is simply a Java program that is embedded in some other software or operating environment. By far the most common use of Java applets is the embedding of applets in web pages on the internet. It should also be borne in mind however that Java is designed to be a platform independent language, and it is possible for applets to be embedded in a wide range of environments ranging from Smart Cards to mobile phones to car navigation systems. In this chapter, however, we shall stick to applets embedded in web pages.

Many Java programming courses and books tend to begin with programming applets, so why have we waited until this chapter to cover them? There are several reasons for this:

- Applets require some form of graphical user interface programming to be useful, which in turn requires some understanding of Java object-oriented programming techniques.

- The most 'transferable' programming skills (and therefore probably most important) are the more generic concepts covered in the earlier part of this course and not the specific details of GUI programming and applets.

- Java applets are essentially a subset of the more powerful Java *applications*. As you can now write full Java applications, writing Java applets requires virtually no extra work.

- For applets to work in web pages, you need a Java equipped web browser. Both Netscape and Microsoft have (inevitably) been somewhat slower than Sun's JDK in implementing Java in their browser.

The best way of appreciating what an applet looks like and how to program one, is to consider a simple example. The applet we will create results in a web page similar to that shown in Figure 10.1.

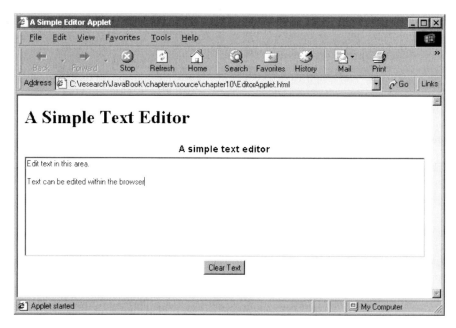

Figure 10.1 Example of a simple text editor applet.

The code to produce this applet is shown below.

```
import java.applet.Applet;      // For the Applet class.
import java.awt.*;              // For other user interface classes.
import java.awt.event.*;        // For action listener.
//   ***************************************************************
/** Creates a simple editable area to embed inside a web page.
  * @author    Jo Wood.
  * @version   1.1, 14th July, 2001
  */
//   ***************************************************************
public class EditorApplet extends Applet implements ActionListener
{
    // -------------------- Object Variables --------------------
```

```java
    private Button bClear;        // Button to clear text.
    private TextArea taEditor;    // Editable text area.

    // -------------------- Applet Methods ---------------------

    /** Displays a editable text area and a button.
      */
    public void init()
    {
        setLayout(new BorderLayout());

        // Title label.
        Label title=new Label("A simple text editor",Label.CENTER);
        title.setFont(new Font("SansSerif",Font.BOLD,14));
        title.setBackground(Color.white);
        add(title, BorderLayout.NORTH);

        // Editable text area.
        taEditor = new TextArea("Edit text in this area.");
        add(taEditor, BorderLayout.CENTER);

        // Clear Button.
        Panel buttonPanel = new Panel();
        buttonPanel.setBackground(Color.white);
        bClear = new Button("Clear Text");
        bClear.addActionListener(this);
        buttonPanel.add(bClear);

        add(buttonPanel,BorderLayout.SOUTH);
    }

    // ------------------ Implemented Methods -------------------

    /** Responds to a mouse click over the button and clears screen
      * when pressed.
      * @param event Mouse event.
      */
    public void actionPerformed(ActionEvent event)
    {
        if (event.getSource() == bClear)
        {
            taEditor.setText("");
        }
    }
}
```

On the whole, the code should look fairly familiar to you as we created a similar class in Chapter 8. You will however notice new aspects to the code. Because we are creating an applet, we inherit the `Applet` class, which does much of the work of communicating with the browser for us. To do this, we need to import the package containing the definition of the `Applet` class. The `Applet` class behaves in much the same way as the `JFrame` class that we have used previously, and it is often possible to substitute one for the other.

You will notice that this class does not have a constructor. Applets can (and often do) have constructors, but they also have another method that is called before the constructor. This method is always called `init()`, and we need to override it to make our `Applet` work (you can think of `init()` as being the equivalent to the `main()` method we have seen in previous application classes). The only other difference between this class and the previous `JFrames` we have created is that there are no `pack()` or `setVisible()` methods. This is because the packing and visibility is handled by the web browser.

10.1.1 Applets and Web Pages

To embed our Java applet in a web page, we have to do two things. First, we must place the Java bytecode in some location that is accessible to a web server. Second, we must create a web page that links to the Java code we have written.

In fact, there is no need to place applet code on a web server if you do not intend to share it with the outside world. It is quite possible to develop your applet locally on your own machine, and link it to a web page on the same machine without any web serving at all. But more usually, the point of writing an applet is to share it over the web by placing it on some server.

A Very Brief Introduction to HTML

In order to embed your Java applet on the web, you need to understand how web pages are created with HTML. If you are unfamiliar with HTML – the Hypertext Markup Language used to create web pages, the following provides a brief introduction.

The HTML language is made up of a series of *tags* that usually surround sections of text. A tag can be identified because it is encased in <triangular brackets>. The tag itself often contains some kind of formatting instruction. Tags usually come in pairs – the first turns some format on, the second, which is identical but with a / after the first bracket, turns it off again. So, for example, to make a word appear in **bold**, the HTML instruction might be:

```
to make a word appear in <b>bold</b>, the HTML instruction
```

where b is the symbol to indicate the emboldening of text. Beyond the formatting of individual blocks of text, all HTML files have the following general structure:

```
<html>
  <head><title>Some title here</title></head>
  <!-- This is a comment and is ignored by the browser. -->
  <body>
    <p> Some text with formatting instructions here.</p>
  </body>
</html>
```

Unlike Java, HTML is not case-sensitive. You can choose to write tags with upper- or lower-case characters. Like Java though, it is useful when writing HTML, to make it as clear as possible through the use of indentation and comments. Notice how it is possible to *nest* tags within each other, so that in the example above, both the `<head>` and `<body>` tags sit within the main `<html>` tag and are indented accordingly. The `<title>` tag in turn sits within the `<head>` tag and the `<p>` tag sits within `</body>`.

Some useful HTML Tags

To create usefully formatted text and images in HTML, you need to know which tags to use and when to use them. The following list describes some of the more commonly used tags.

Commenting

Like Java, you can add comments to HTML that will be ignored by the web browser. Comments are encased within `<!--` and `-->` and can stretch over several lines if required.

Paragraph formatting

Each time you include a paragraph of text in your document, it should be encased in the paragraph tags `<p>` and `</p>`.

If you wish to include headings in your document, you can use the heading tags `<h1>` and `</h1>`. Note that the number after the `h` can be from 1 to 6, where 1 is the largest heading, and 6 is the smallest.

Text styles

The common text formatting tags include `<b>`, `</b>` for **emboldening** text, `<i>`, `</i>` for *italicising* text and `<u>`, `</u>` for underlining text.

Hyperlinks

To create a *Hyperlink* to another web page the syntax is `<a href="url"> some hyperlinked text</a>`, where `url` the URL of the web page with which you wish to link. For example,
`<a href="http://www.soi.city.ac.uk/~jwo/index.html">Home</a>`

Images

To display a graphics image in a web page (which should be in either the GIF or JPEG format, the tag to use is `<img src="imageFile" width="width" height="height"`, where `imageFile` is the file or URL containing the image and `width` and `height` are the dimensions of the image as it is to appear on the page.

This introduction barely scrapes the surface of HTML, but should be enough to allow you to embed your own Java applets in web pages.

An example of HTML code that links the applet above to a web page is shown below. It provides a template that can be used to call any applet from within a web page.

```
<!DOCTYPE HTML PUBLIC "-//W3C//DTD HTML 4.0 Transitional//EN">

<html>
 <head>
  <title>A Simple Applet Template</title>
 </head>
 <!--- End of Header --->

 <body bgcolor="white">

  <h1>Simple Applet Template</h1>

  <applet code="EditorApplet.class" codebase="editor"
          width="600" height="200">
   <p>
     <i>This browser is not running Java Applets</i>
   </p>
  </applet>

 </body>

</html>
```

The simplest HTML instruction for linking two applets is the `<applet>` tag. The most useful attribute of this tag is `code`, which specifies the name of the Java bytecode to embed. Note that unlike running Java applications, you must specify the `.class` extension to this file as well as the class name. If the bytecode does not sit in the same directory as the HTML file calling it, you must also specify its location using `codebase`. In the example above, the bytecode is in a sub-directory called `editor`. Finally, you must specify the amount of space on the web page given over to the applet using the `width` and `height` attributes.

If you place your Java applet on the web, you cannot be sure which platforms and which browsers might browse your page. It may be that the page is displayed by a browser which cannot or does not run Java applets (see below on why this might be the case). To cover such eventualities, it is useful to place some HTML code for those that cannot display applets. Any code between the `<applet>` and `</applet>` tags will be ignored if the browser can run applets, but displayed if does not.

10.1.2 Passing Parameters to Applets

Suppose you have written an applet that you would like others to be able to embed in their own web pages. There may be certain aspects of this applet that you would like others to be able to modify. However, with what we have seen so far, in order to modify an applet, someone would have to have access to the Java source code and a knowledge of the Java language.

An alternative approach is to allow *parameters* to be passed from the web page into your Java applet. Thus, others are able to customise your applet without being exposed to the Java code directly. Using parameters involves two steps. First, the parameter names and values must be specified inside the HTML `<applet>` tag. Second, your Java applet must extract these parameters from the HTML and incorporate them into your Java code.

To illustrate the process, consider how we might modify our `Editor` applet to display text sent to as an applet parameter. Our first step is to modify the HTML that calls the applet:

```
<applet code="EditorApplet.class" codebase="editor"
        width="600" height="200">
  <param name="initText"
         value="This text was taken from the HTML page">
</applet>
```

Using the `<param>` tag, the HTML passes a single parameter called `initText` to the applet. The content of the parameter is given by the `value` attribute.

The next stage is to get Java to extract the parameter information and do something with it. This is done with the method inside the `Applet` class called `getParameter()`. This method takes a `String` as an argument that represents the name of the parameter to grab, and returns another `String` containing the contents of that parameter.

So, for example, to extract contents of the `initText` parameter and initialise our editor with its value, we would use the following two lines of code:

```
String initialText = getParameter("initText");
taEditor = new TextArea(initialText);
```

10.1.3 Applet Security

The major advantage of producing Java web applets as opposed to Java applications is that once on a public web page, your program is accessible to the outside world. Anyone with a Java enabled web browser can run your program. They do not need to have the SDK on their system as the browser does all the Java interpretation.

That same advantage comes with a big restriction on your programming too. It would be possible to write a Java application that erased every file on the hard disk of your computer. It is assumed that no one would be unwise enough to write such a program on their own machine, but if such programs could be run by simply visiting a web page, the consequences for worldwide distributed computing would be serious. Consequently, applets have a set of built-in security restrictions that stop such activity. The exact nature of the security restrictions may vary between browsers and even different versions of the same browser. As both Netscape and Microsoft Internet Explorer have evolved, they have tended to become stricter in what they allow a Java applet to do. In particular, you are unlikely to be able to do the following with an applet downloaded from the web:

- Read a file on the local machine.
- Write a file on the local machine.
- Connect via the *port* of the local machine.
- Identify any directories or resources on the local machine.

The restrictions above ensure that any applets you find (or create) on the web are unlikely to do any damage to your machine, or reveal any critical information about your machine to the outside world. Note, that these restrictions may be circumvented in special cases where a so-called *trusted applet* is created. However, the procedure for doing this is complicated, and we will not dwell on it here.

So do these restrictions stop us writing useful applets? Well, almost, but the major exception to the 'file embargo' is that the applet may read files from the same directory that applet was downloaded from (i.e. on the server). It may also gather input from the keyboard of the client, as the simple text editor above demonstrated.

10.1.4 Applet Browser Restrictions

The other main restriction you will notice when using applets in web browsers is that some Java methods and classes will not be recognised by the browser. In particular, you may find that none of the Swing classes can be used in applets that have been embedded as above. To understand why this is the case, you need to consider how *Java Virtual Machines* are developed.

When a new version of Java is released by the main developers of Java (Sun), they will release a Software Development Kit that allows Java programs to be compiled and interpreted. This is what is used every time you compile and run an application using `java`, `javac` or possibly an Integrated Development Environment. The developers will also release an *Application Programming Interface* or API for the latest version of the language. This is a specification of the methods and classes that are used by the system, the documentation for which is given in the familiar `javadoc` style.

For a program to work in a browser, the browser developer (for example, Microsoft or Netscape) has to write their own *Virtual Machine* to interpret the Java bytecode in the browser. This can only be done once the API has been specified by Sun. The result is that the version of Java understood by a browser is likely to lag behind the latest version released by Sun.

Microsoft Internet Explorer prior to versions released with Windows XP, and most versions Netscape can interpret Java 1.1 or below using their in-built *Virtual Machines*. However, they tend to have problems with code specific to Java 2. If you wish your Java applets to be readable by the majority of web browsers, you should limit your code to that which is part of the Java 1.1 release. Table 10.1 shows some of the more common classes and packages that should be avoided if you wish to make your applets available to all browsers. For a more thorough list, the API documentation for all Java classes should indicate if a particular class or package was introduced after Java 1.1.

Table 10.1 Some classes to avoid when creating Java 1.1 applets

Package	Alternative
`javax.swing`	Swing classes should be replaced with their AWT equivalents. For many classes this simply means removing the 'J' prefix (e.g. `JFrame` to `Frame`, `JButton` to `Button`, `JMenu` to `Menu`). Other more sophisticated Swing classes have no direct equivalent (e.g. `JFormattedTextField`) and must be programmed 'by hand'.
`java.util.Collection`	The `Collection` interface for dynamic collections (e.g. `HashMap`, `LinkedList`). Dynamic collections should be limited to those released with Java 1.1 (`Vector`, `Hashtable` and `Dictionary` the most common). The `Iterator` class should be replaced with `Enumerator`.
`java.awt.geom`	Java2D classes such as `GeneralPath`, `Stroke` and `AffineTransform` should be replaced with their AWT equivalents.

There is a solution that allows any browser to use the latest version of Java, known as the *Java plug-in*. This technology downloads the latest Java Virtual Machine from Sun and links it to your web browser. The main advantage of the Java plug-in, is that it guarantees that the client browser has an up-to-date version of the Java Virtual Machine. The main disadvantage of this process is that many users will not have downloaded the plug-in, so your applet is likely to be available to less people than one written in Java 1.1. However, the trend in web browser development, illustrated by the release of Microsoft's Internet Explorer with Windows XP, and Netscape 6 suggests that the plug-in route will become increasingly widely used in the future. The procedure for linking an applet to a plug-in version of the Java virtual Machine is a little more complicated than HTML described above and will not be discussed in any great detail in this book. For more details, see the case study at the end of this chapter and Sun's Java Plug-in page `java.sun.com/products/plugin`

Testing an Applet with the AppletViewer

The Java SDK comes with a tool specifically designed for testing Java applets. This allows you to run applet code without depending on a Java-enabled browser. It uses the same Java Virtual Machine as the SDK so is guaranteed to be up-to-date.

To start the *AppletViewer*, issue the following command either from a shell (Unix or DOS), or from the `Run...` command in the Windows `Start` menu.

```
appletviewer fileName.html
```

where *fileName*.html is the HTML page linking to your applet. The AppletViewer will ignore any HTML other than that placed in the `<applet>` tags. Output from the AppletViewer using the text editor applet given in Section 10.1.1 is shown below.

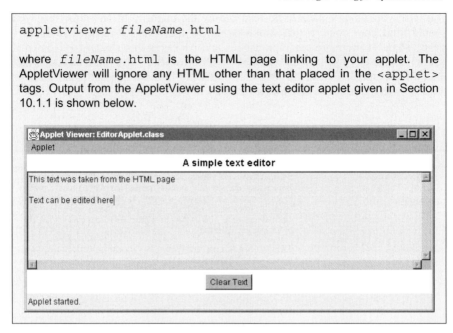

10.2 CASE STUDY: A SIMPLE MAP APPLET

Suppose we wish to create an applet that will display a map in a web browser such that when a user moves the mouse over any location on the map, it reports the georeferenced grid coordinates of the location. This result is similar to that achieved with the `ModelDisplay` class developed in the previous chapter. However, since we wish to embed this class in a web page, we will have to remove the file input/output functionality and restrict code to Java 1.1.

The first stage of the process is to create an applet that extracts the name of an image file representing the map. The `Applet` class has a method that can read both JPEG and GIF images. This method returns an object of type `Image` that stores the graphical information. We have seen that security restrictions mean that applets can only read files from the same location as the Applet itself. To find out this location, we can use another method from the `Applet` class called `getCodeBase()`.

So, to read an image file from the applet location and store it in an `Image` object, we need to include a line similar to the following:

```
Image map = getImage(getCodeBase(),getParameter("fileName"));
```

where `getParameter("fileName")` attempts to extract the name of the file from the applet.

We also need to extract the geographical boundaries associated with the image. These are likely to be different to the pixel dimensions of the image so they must be extracted explicitly. We can maximise the flexibility of our program by extracting these as applet parameters.

```
try
{
    north = new Integer(getParameter("north")).intValue();
    south = new Integer(getParameter("south")).intValue();
    east  = new Integer(getParameter("east")).intValue();
    west  = new Integer(getParameter("west")).intValue();
}
catch (NumberFormatException e)
{
    // Use default values if problem extracting boundaries.
    north = 1;
    south = 0;
    east  = 1;
    west  = 0;
}
```

The `getParameter()` method always returns a `String` representing the parameter, even if it represents a number. Consequently, we need to convert each parameter into an integer representing each boundary. As it is possible that the parameter values could be missing or given incorrectly (something we as Java programmers have no control over), we use a `try-catch` combination to anticipate any conversion errors.

The next task is to display the map image in the applet. In keeping with the object-oriented philosophy, we will create a new class specifically designed for displaying images called `ImagePanel` just as we did for `RasterMap` and `VectorMap`.

```
import java.awt.*;         // Required for graphical objects.

// ***************************************************************
/** Panel that displays an image.
  * @author    Jo Wood.
  * @version   1.0 5th August, 2001
  */
// ***************************************************************

public class ImagePanel extends Panel
{
    // ------------------- Object Variables ------------------

    private Image image;

    // -------------------- Constructor --------------------

    /** Creates an image panel that will display the given Image.
      * @param image Image to display in panel.
      */
    public ImagePanel(Image image)
```

```
    {
        super();
        this.image = image;      // Store image as object variable.
    }

    // ----------------- Overridden methods ----------------

    /** Displays the image in the panel.
      * @param g Graphics context in which to paint.
      */
    public void paint(Graphics g)
    {
        // Find size of the panel and stretch image over it.
        Dimension size   = getSize();

        // Draw the image with a simple border.
        g.drawImage(image,0,0, size.width, size.height,this);
        g.drawRect(0,0, size.width-1, size.height-1);
    }
}
```

The class inherits the AWT class Panel so that it can be combined with other graphical objects just as a Panel can. Notice that we are not using the Swing class JPanel as we are keeping the code Java 1.1 compliant to work with the maximum number of browsers. For the same reason, we customise the drawing behaviour of the panel overriding the AWT paint() method rather than the Swing paintComponent() method.

The constructor calls Panel's normal constructor and then additionally stores the given Image object. The display is achieved by overriding the paint() method that is present in all AWT Components. This method finds out how large the panel is and then calls the method drawImage() to display the image in the panel. drawImage takes several parameters that allow the position of the image, its drawing dimensions and the name of the component doing the drawing to be specified. Finally, the method draws a rectangle around the image using the method drawRect() that comes with the Graphics class.

The ImagePanel class can store and display any image. To make our applet display a particular image, we need to create an ImagePanel object and add it to the applet:

```
mapPanel = new ImagePanel(map);
add(mapPanel,BorderLayout.CENTER);
```

Finally, we must allow the user to interrogate the image with the mouse using a MouseMotionListener just as we did for SpatialPanel in Chapter 8.

All that remains is to convert the mouse-based pixel coordinates into georeferenced values based on the geographical boundaries of the image. Unfortunately, the AffineTransform class that we have previously used to do this is not available in Java 1.1, so we have to perform the scaling 'by hand'. The conversion requires us to know the size of the image and the geographical bounds of the map:

```
easting  = ((float)x/imageWidth)*(west-east) + west;
northing =((float)(imageHeight-y)/imageHeight)*(north-south) + south;
```

Note that the pixel coordinate system has the origin in the top-left corner whereas geographical referencing uses the bottom-left as origin. Consequently, we have to flip the y-axis as part of the transformation. We also typecast the pixel coordinates as floating point numbers to ensure precision of the division calculation.

We now have enough information to assemble the final applet:

```java
import java.awt.*;         // Required for graphical objects.
import java.awt.event.*;   // For mouse listening.
import java.applet.*;      // For the Applet class.

// ****************************************************************
/** Displays map on screen and allows simple mouse interrogation.
 *  @author    Jo Wood.
 *  @version   1.2 5th August, 2001
 */
// ****************************************************************

public class MapInterrogator extends Applet
{
    // ------------------- Object Variables --------------------

    private ImagePanel mapPanel;            // Map to display.
    private int north,south,east,west;      // Georeferenced map bounds.
    private Label coords;                   // Displays position.


    // ---------------------- Methods ------------------------

    /** Reads in the graphics file ready to display.
      */
    public void init()
    {
        // Set layout manager and add title to top.
        setLayout(new BorderLayout());
        add(new Label("Map Display"),BorderLayout.NORTH);

        // Store the georeferenced map boundaries.
        try
        {
            north = new Integer(getParameter("north")).intValue();
            south = new Integer(getParameter("south")).intValue();
            east  = new Integer(getParameter("east")).intValue();
            west  = new Integer(getParameter("west")).intValue();
        }
        catch (NumberFormatException e)
        {
            // Use default values if problem extracting boundaries.
            north = 1;
            south = 0;
            east  = 1;
            west  = 0;
        }
```

```
        // Transfer graphics file into an Image object.
        Image map=getImage(getCodeBase(),getParameter("fileName"));

        // Create and display mouse listening image map.
        mapPanel = new ImagePanel(map);
        mapPanel.setCursor(new Cursor(Cursor.CROSSHAIR_CURSOR));
        mapPanel.addMouseMotionListener(new MouseMoveMonitor());
        add(mapPanel,BorderLayout.CENTER);

        // Add label describing mouse position.
        coords = new Label("Move mouse over map for location...");
        add(coords,BorderLayout.SOUTH);
    }

    // --------------------- Nested Classes ---------------------

    /** Handles mouse movement over the panel.
      */
    private class MouseMoveMonitor extends MouseMotionAdapter
    {
        /** Handles mouse movement over the panel by reporting the
          * coordinates of the current mouse position.
          * @param e Mouse event associated with movement.
          */
        public void mouseMoved(MouseEvent e)
        {
            Dimension imageSize = mapPanel.getSize();
            float easting  = ((float)e.getX() / imageSize.width) *
                              (east-west) + west;
            float northing = ((float)(imageSize.height-e.getY()) /
                              imageSize.height)*(north-south)+south;
            coords.setText("Location: "+ easting+" , "+northing);
        }
    }
}
```

The working applet and the HTML code required to invoke it are shown below.

```
<applet code="MapInterrogator.class"
        width="525" height="450"
        codebase="MapInterrogator">

  <param name="fileName" value="map.jpg">
  <param name="north" value="3500">
  <param name="south" value="0">
  <param name="east" value="4000">
  <param name="west" value="0">
  <i>This browser is not running Java Applets</i>

</applet>
```

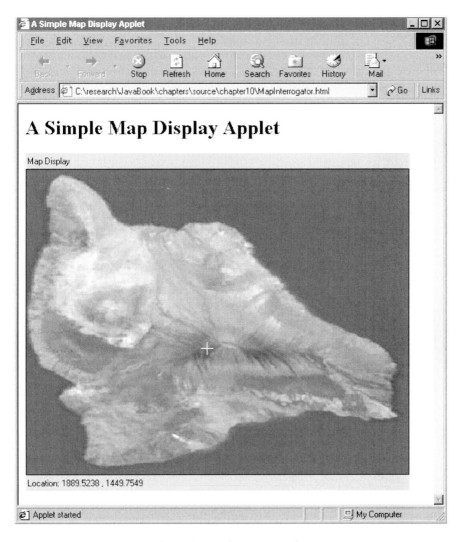

Figure 10.2 Map interrogator applet.

10.3 COMMUNICATING USING XML

Once of the most interesting developments in internet-based communication in recent years has been in the use of the extensible markup language XML. XML allows users to create their own markup languages and documents that may be shared over the internet. Unlike HTML, XML allows users to create *semantic* markup that describes the meaning of elements within a document rather than the way they should be displayed.

XML for Beginners

It is not within the scope of this book to describe how XML works in any detail, but if you have not used XML and wish to learn more about the language, you should consult one of the references given at the end of this book.

XML and its associated technologies such as XSL, XSLT, DTD and schemas can become somewhat complicated and bewildering. The key concepts you should understand given below along with a simple example XML file.

```
<?xml version="1.0"?>

<!-- An example of a spatial XML file -->

<map>

 <title>An example of a simple map</title>

 <polygon name="lake">
  <coordinates>10,35 6,25 20,11</coordinates>
 </polygon>

 <line name="Minor road">
  <coordinates>30,45 80,55</coordinates>
 </line>

</map>
```

- An XML file consists of *tags* enclosed in <triangular> brackets. For every opening <tag> there is usually an equivalent closing </tag>. A pair of tags identifies an XML *element*.
- Tags may be nested within one another and therefore by implication, XML elements can contain other elements.
- Tags can also have optional *attributes* that are given a name and value within the opening tag. For example, name is an attribute of both the line and polygon tags in the example above).
- *Comments* can be created by enclosing text within <!-- and -->.
- Unlike HTML, XML allows you to create your own tags and attribute types. The types of tag available are often declared in a separate file called a *document type definition* (DTD).
- Unlike HTML, XML documents must be *well formed*. That is they must conform to certain grammatical rules such as matching opening and closing tags, attribute values enclosed in quotes etc.
- If an XML file is linked to a DTD, it should also be *valid*. That is, should only contain the types of elements and attributes identified by the DTD, and should only be structured according to the rules in the DTD.

You can both read and write XML files in Java. This can be useful when you wish to write a Java program that performs some action in response to the contents of an XML file, or you wish to export the contents of some data structure in a form readable by other applications.

10.3.1 Setting Up Java to Handle XML

In order to manipulate XML files in Java, you need to do two things. First, you need to be able to link your own programs with an *XML parser*. That is, a program that when presented with an XML file, can identify its individual elements and attributes. There are many of these freely available such as the *Xerces Parser* produced by Apache.

Second, you need some way to interface your code with the XML language. The most widely used interface is called the *SAX* (Simple Api for Xml). This provides you with a series of classes and methods for linking to an XML parser and extracting the elements and attributes from an XML file.

Prior to version 1.4 of Java from Sun, neither SAX nor an XML parser was included with the standard language installation. However, Sun provide a single download called JAXP that contains all you need to write your own XML manipulation programs. For versions of Java earlier than 1.4, you will have to install this package separately on your system. You can find instructions on how to install JAXP in the references section at the end of this book.

Assuming JAXP has been installed on your system, any class that you create that wishes to use handle XML files will have to import the relevant packages:

```
import javax.xml.parsers.*;
import org.xml.sax.*;
import org.w3c.dom.*;
```

You will then need to establish a link with an XML parser. The easiest way of doing this is to get JAXP to establish a link using what is known as a DocumentBuilderFactory. In Java programming terms, a *factory* is a class designed to create a range of other types of object. It is used so that platform-specific code can be separated from more generic cross-platform class design.

```
DocumentBuilderFactory dbf = DocumentBuilderFactory.newInstance();
DocumentBuilder db=null;
try
{
    db = dbf.newDocumentBuilder();
}
catch (ParserConfigurationException e)
{
    System.out.println("Problem finding an XML parser: "+e);
}
```

In this case, the factory creates an XML parser based on the setup on your particular machine. This allows you (or any user of your Java program) to 'plug-in' a third party XML parser into their system. The factory will then detect this local implementation and allow your Java code to use it. At this stage, we will not worry about how to use different parsers since JAXP comes with its own XML parser available automatically with the JAXP package.

10.3.2 The Document Object Model (DOM)

Once we have called an XML parser, there are several ways to store the contents of an XML file. The one we shall consider here is known as the *Document Object Model* or DOM. A DOM consists of a series of objects linked in an hierarchical tree. Each node of the tree represents an object such as an XML element or attribute. The DOM tree for the sample XML file shown in 'XML for Beginners' above is represented graphically in Figure 10.3.

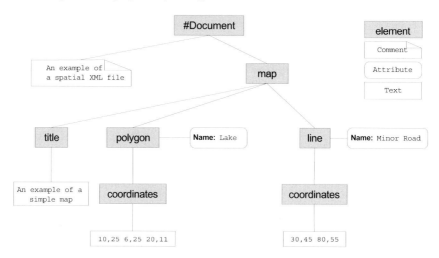

Figure 10.3 A DOM tree structure of a simple XML file.

The code to create a DOM from an XML file is shown below:

```
Document dom=null;
try
{
   dom = db.parse(file);
}
catch (SAXException e)
{
   System.err.println("Problem parsing document: "+e.getMessage());
}
```

where `file` is an object of type `File` that represents the XML file to read and the class `Document` stores the entire DOM.

Navigating a DOM can seem a little complicated at first since a DOM tree can consist of many branches and nodes. Any one node in a DOM is represented using the `Node` class which contains a variety of useful methods for identifying the type, name and contents of a node. The easiest way of understanding how to navigate the nodes of a DOM is to consider the following example, which simply reads a given XML file and prints out its structure node by node:

```
package jwo.jpss.utilities;    // Part of the JPSS utilities package.
import javax.xml.parsers.*;    // For Document builder factory.
import org.xml.sax.*;          // For SAX Exception handling.
import org.w3c.dom.*;          // For document object model (DOM).
import java.io.*;              // For file handling.

// ****************************************************************
/** Reads and writes XML files.
  * Uses the Document Object Model (DOM) to store the tree of
  * nodes represented by the XML file.
  * @author Jo Wood.
  * @version V1.3 29th August, 2001.*/
// ****************************************************************
public class XMLProcessor
{
    // -------------------- Object Variables --------------------

    private int indent;              // Indent level.
    private Document dom;            // Document object model.
    private PrintWriter out;         // Output stream.

    // --------------------- Constructors ----------------------

    /** Creates an XML processor using the given DOM.
      * @param DOM to use in processor.
      */
    public XMLProcessor(Document dom)
    {
        this.dom = dom;
    }

    /** Reads the given XML file and constructs a DOM from it.
      * @param fileName Name of XML file to read.
      */
    public XMLProcessor(String fileName)
    {
        readXML(fileName);
    }

    // ---------------------- Methods ------------------------

    /** Reads the given XML file and converts it into a DOM.
      * @param fileName Name of XML file to convert.
      * @return True if converted successfully.
      */
```

```java
public boolean readXML(String fileName)
{
    DocumentBuilderFactory dbf =
                        DocumentBuilderFactory.newInstance();
    DocumentBuilder db = null;
    indent = 0;

    try
    {
        db = dbf.newDocumentBuilder();
    }
    catch (ParserConfigurationException e)
    {
        System.out.println("Problem finding XML parser:\n"+e);
        return false;
    }

    // Try to parse given file and store XML nodes in the DOM.
    try
    {
        dom = db.parse(new File(fileName));
    }
    catch (SAXException e)
    {
        System.out.println("Problem parsing document: "+e);
        dom = db.newDocument();
        return false;
    }
    catch (IOException e)
    {
        System.out.println("Problem reading "+fileName);
        return false;
    }
    return true;
}

/** Displays the DOM stored within this class as an XML file
  * using standard output.
  */
public boolean writeXML()
{
    out = new PrintWriter(System.out);
    indent = 0;
    outputNodeAsXML(dom);
    out.close();
    return true;
}

/** Converts the DOM stored within this class into an XML file.
  * @param fileName Name of file to contain the XML.
  * @return true if successful XML generation.
  */
public boolean writeXML(String fileName)
{
    if (dom == null)
    {
        System.out.println("Error: No DOM to process.");
        return false;
```

```java
        }

        // Open file for output.
        try
        {
            out = new PrintWriter(new BufferedWriter(
                                  new FileWriter(fileName)));
        }
        catch (IOException e)
        {
            System.out.println("Error: Problem creating XML file");
            return false;
        }

        // Start recursive output of the whole DOM.
        indent = 0;
        outputNodeAsXML(dom);

        // Close output and leave.
        out.close();
        return true;
    }

    // -------------------- Private Methods ---------------------

    /** Converts the given DOM node into XML. Recursively converts
     *  any child nodes.
     *  @param node DOM Node to display.
     */
    private void outputNodeAsXML(Node node)
    {
        // Store node name, type and value.
        String name  = node.getNodeName(),
               value = node.getNodeValue();
        int    type  = node.getNodeType();

        // Ignore empty nodes (e.g. blank lines etc.)
        if ((node==null) || ((value != null) &&
            (value.trim().equals(""))))
            return;

        switch (type)
        {
            case Node.DOCUMENT_NODE:   // Start of document.
            {
                out.println("<?xml version=\"1.0\" ?>");

                // Output the document's child nodes.
                NodeList children = node.getChildNodes();

                for (int i=0; i<children.getLength(); i++)
                    outputNodeAsXML(children.item(i));
                break;
            }

            case Node.ELEMENT_NODE:    // Element with attributes.
            {
                // Output opening element tag.
```

```java
                    indent++;
                    indent();
                    out.print("<"+name);

                    // Output any attributes the element might have.
                    NamedNodeMap attributes = node.getAttributes();
                    for (int i=0; i<attributes.getLength(); i++)
                    {
                        Node attribute = attributes.item(i);
                        out.print(" "+attribute.getNodeName()+
                                  "=\""+attribute.getNodeValue()+"\"");
                    }
                    out.print(">");

                    // Output any child nodes that exist.
                    NodeList children = node.getChildNodes();

                    for (int i=0; i<children.getLength(); i++)
                        outputNodeAsXML(children.item(i));

                    break;
                }

                case Node.CDATA_SECTION_NODE:   // Display text.
                case Node.TEXT_NODE:
                {
                    out.print(value);
                    break;
                }

                case Node.COMMENT_NODE:         // Comment node.
                {
                    indent++;
                    indent();
                        out.print("<!--"+value+"-->");
                    indent--;
                        break;
                }
            }

            // Finally output closing tags for each element.
            if (type == Node.ELEMENT_NODE)
            {
                out.print("</"+node.getNodeName()+">");
                indent--;
                indent();
            }
    }

    /** Indents output to current tree depth.
      */
    private void indent()
    {
        out.println();
        for (int i=1; i<indent; i++)
            out.print(" ");
    }
}
```

Communicating with the Wider World   287

Output from this class is controlled by the methods `writeXML()` and `writeXML(String)`. The former uses standard output (to the screen), while the latter allows output to be sent to a file. Rather than writing two separate methods for issuing output, the output stream is represented as a `PrintWriter`. This class can be used to initiate an output stream as either standard output (`System.out`) or file output (`FileWriter`).

The method `outputNodeAsXML()` firstly identifies the name, type and contents of the node using the methods `getNodeName()`, `getNodeType()` and `getNodeValue()`. This information can be used to display either the contents of text, comment and attribute nodes (see unshaded boxes in Figure 10.3) or the name of element nodes (shaded boxes in Figure 10.3).

In all but the simplest trees, an element node is likely to contain further sub-nodes within it. The last part of the `outputNodeAsXML()` looks for these 'child' nodes and recursively calls itself, this time displaying the contents of the child node. Thus, the method traverses the entire tree until all nodes have been displayed.

We can create an application that processes an XML file by creating a class with a main method. In the example below, we can extract and the name of the file to process from any command line arguments supplied to the application. If none is provided, the default name of `text.xml` is used.

```
import jwo.jpss.utilities.*;     // For XMLProcessor.

// ****************************************************************
/** Starts the XMLProcessor with a file name given by command
 *  line arguments.
 *  @author Jo Wood.
 *  @version V1.2 12th August, 2001. */
// ****************************************************************

public class RunProcessor
{
    // -------------------- Main Method ----------------------

    public static void main(String[] args)
    {
        // Extract name of XML file to process (or use default
        // if not specified).
        String fileName;

        if (args.length != 1)
            fileName = new String("test.xml");
        else
            fileName = args[0];

        // Create an XML processor.
        XMLProcessor xmlProcessor = new XMLProcessor(fileName);
        xmlProcessor.writeXML();
    }
}
```

While this example has been relatively simple, it illustrates the process of extracting the individual elements that are represented by an XML file. Of more use would be a method that searched for a given element and returned the text encased by it. An XML file might contain multiple instances of the same element (e.g. `<coordinates>` in the example file), so such a search method would have to return a collection of text items.

```
private Vector matches;

/** Searches for a given element and returns all text nodes
  * associated with it.
  * @param element Element to search for.
  * @return Array of strings associated with all occurrences
  * of the given element.
  */
public String[] search(String element)
{
    // Search for text nodes associated with element.
    matches = new Vector();
    search(dom,element);

    // Convert match vector into an array.
    String[] matchArray =  new String[matches.size()];
    matches.toArray(matchArray);
    return matchArray;
}
```

This public method initialises a `Vector` of `Strings` for storing matched text, then calls a private search method to traverse the DOM. After completing the search, the `Vector` is converted into an array to be returned as a message.

The private search method traverses the DOM in much the same way as `outputNodeAsXML()`. In this case, the method is simpler as it only searches for text nodes and elements with children.

```
/** Searches for a given element and updates list of matching text.
  * @param node Node to start search from.
  * @param element Element to search for.
  */
private void search(Node node, String element)
{
    if (node.getNodeName().equalsIgnoreCase(element))
    {
        // Match found, so look for text in children.
        NodeList children = node.getChildNodes();

        for (int i=0; i<children.getLength(); i++)
        {
            Node child = children.item(i);

            if ((child.getNodeType() == Node.CDATA_SECTION_NODE) ||
                (child.getNodeType() == Node.TEXT_NODE))
                matches.add(child.getNodeValue());
        }
    }
```

```
    if ((node.getNodeType() == Node.DOCUMENT_NODE) ||
        (node.getNodeType() == Node.ELEMENT_NODE))
    {
        // Search child nodes.
        NodeList children = node.getChildNodes();

        for (int i=0; i<children.getLength(); i++)
            search(children.item(i),element);
    }
}
```

Calling search("coordinates") using the example XML file, will return an array of two Strings

```
10,35  6,25,  20,11
30,45  80,55
```

These can be used in the construction or display of new spatial models.

10.4 CASE STUDY: CONVERTING APPLICATIONS INTO APPLETS

In this final case study, we will consider how we can convert our two major application developments – the ant simulation and the spatial model display into web-friendly applets. Each application conversion demonstrates a different approach to creating applets. The first adapts the existing (ant simulation) code so that it becomes compatible with Java 1.1 web browsers. The second shows how the Java plug-in can be used to force the browser to use a more recent version of the Java Virtual Machine.

10.4.1 Creating a Java 1.1 Ant Simulation Applet

Let us assume that we wish to share the ant simulator developed in previous chapters with as many potential users as possible. Turning the simulation into an applet seems like an obvious approach, especially as the simulation does not involve any file input/output – one of the sets of operations not permitted by most applets. To maximise the potential audience for the applet, we will attempt to make the simulation Java 1.1 compliant.

Table 10.1 summarised some of the more important classes that were introduced with Java 2 and which should therefore be avoided in our applet classes. In particular, the ant simulation uses the following Java 2 classes and methods.

- Swing classes, JFrame, JPanel and methods paintComponent(), paintImmediately() (in AntApplication and GardenPanel).
- Collection class Iterator and methods add(), addAll() and remove() (in Garden, Nest, Ant and Queen).

One way of identifying classes not included in a browser's Virtual Machine is simply to run the applet and look for errors. Runtime errors such as these will be displayed in the browser's 'Java Console', usually accessable as a menu item in the browser. For example, running the unmodified applet produces the following error in Internet Explorer's Java console window.

```
java.lang.NoSuchMethodError:
          java/util/Vector: method add(Ljava/lang/Object;)Z not found
     at jwo/jpss/ants/Garden.<init> (Garden.java)
     at jwo/jpss/ants/AntApplet.init (AntApplet.java:37)
     at com/ms/applet/AppletPanel.securedCall0 (AppletPanel.java)
     at com/ms/applet/AppletPanel.securedCall (AppletPanel.java)
     at com/ms/applet/AppletPanel.processSentEvent (AppletPanel.java)
     at com/ms/applet/AppletPanel.processSentEvent (AppletPanel.java)
     at com/ms/applet/AppletPanel.run (AppletPanel.java)
     at java/lang/Thread.run (Thread.java)
```

This indicates that the method `add()` used by a Java `Vector` class was not found. This method was introduced with Java 2 in an attempt to standardise the way in which dynamic collections are handled (see Chapter 7). Its Java 1.1 equivalent is the slightly longer (but identical in function) method `addElement()`. Similar changes need to be made to the `Vector`'s iterators, replacing `Iterator` with the Java 1.1 equivalent `Enumeration`. An example of the change is shown below.

Java 2 iteration code (from `Nest`*)*

```
// Let each ant in the nest go about its business.
Iterator i = ants.iterator();
if (i.hasNext())
{
    do
    {
        Ant ant = (Ant)i.next();
        ant.evolve();

        // Remove any dead ants from the colony.
        if (!(ant.isAlive()))
            i.remove();
    }
    while (i.hasNext());
}
```

Java 1.1 iteration code (from `Nest`*)*

```
// Let each ant in the nest go about its business.
Enumeration enum = ants.elements();
if (enum.hasMoreElements())
{
    do
    {
        Ant ant = (Ant)enum.nextElement();
```

```
        ant.evolve();

        // Remove any dead ants from the colony.
        if (!(ant.isAlive()))
            ants.removeElement(ant);
    }
    while (enum.hasMoreElements());
}
```

The remaining conversion code requires us to create an AWT-based GUI rather than the `JFrame` and `JPanel` used by the Java 2 application. This is not a major task as we have been careful to separate the GUI code from the rest of the simulation. We simply need to replace the `JFrame`-based `AntApplication` with an `Applet`-based `AntApplet`.

```
package jwo.jpss.ants;        // Part of the ant simulation package.
import java.applet.Applet;    // Required for applet functionality.
import java.awt.*;            // For GUI.
import jwo.jpss.spatial.*;    // For spatial classes.
import java.util.*;           // For Java vector class.
// ****************************************************************
/** Applet front end to the Ants in the Garden simulator.
  * @author   Jo Wood.
  * @version  1.2, 26th July, 2001
  */
// ****************************************************************
public class AntApplet extends Applet
                   implements GraphicsListener, Runnable
{
    // ------------------- Object variables ---------------------

    private Thread evolution;       // Evolutionary process.
    private Garden garden;          // Garden containing ants.

    private Image offscreenImage;   // Double-buffering objects.
    private Graphics offscreenGraphics;

    // -------------------- Starter method ---------------------

    /** Initialises applet within which the ants may be observed.
      */
    public void init()
    {
        evolution = new Thread(this);
        Dimension appletSize = getSize();

        // Create an offscreen image and garden on which to draw.
        offscreenImage= createImage(appletSize.width,
                                    appletSize.height);
        offscreenGraphics = offscreenImage.getGraphics();
        garden = new Garden(new Footprint(0,0,
                            appletSize.width,appletSize.height));
```

```java
        garden.addGraphicsListener(this);
    }

    /** Starts off the ant simulation.
     */
    public void start()
    {
        evolution.start();
    }


    // --------------------- Methods ----------------------

    /** Allows drawing in this applet using double buffering.
     *  @param g Graphics context within which to draw.
     */
    public void paint(Graphics g)
    {
        garden.paint(offscreenGraphics);
        g.drawImage(offscreenImage,1,1,null);
    }


    // ---------------- Implemented methods ----------------

    /** Controls the evolution of the simulation as a threaded
     *  process.
     */
    public void run()
    {
        garden.startEvolution();
    }

    /** Stops the ant simulation if window is closed.
     */
    public void stop()
    {
        garden.stopEvolution();
        showStatus("Ant simulator stopped");
        super.stop();
    }

    /** Redraws any graphics that need updating.
     */
    public void redrawGraphics()
    {
        Graphics g = getGraphics();

        if (g != null)
            paint(g);
    }

    /** Checks that given spatial object can be drawn in applet.
     *  @param spObject Spatial object we wish to draw.
     *  @return True if the spatial object can be drawn.
     */
    public boolean canDraw(SpatialObject spObject)
    {
```

```
            if (garden.compare(spObject) == SpatialModel.ENCLOSES)
                return true;
            else
                return false;
        }

        /** Reports list of SpatialObjects associated with the given
          * spatial object (within, matching, etc.).
          * @param spObject spatial object with which to compare.
          * @return List of connected spatial objects.
          */
        public Vector objectsAt(SpatialObject spObject)
        {
            return garden.objectsAt(spObject);
        }
}
```

In addition to inheriting `Applet` rather than `JFrame`, two changes have been made to this container class. First, the process of dynamic evolution has been placed in its own thread. This is generally good practice when creating computationally demanding processes in an applet. This allows the browser to go about its normal business of managing the applet while at the same time allowing the applet to run. Since `AntApplet` already inherits `Applet`, we have to implement the `Runnable` interface and create a `Thread` specifically for handling evolution (see Chapter 8).

The second change concerns the display of the simulation. In the original `AntApplication`, we created a special `GardenPanel` (inherited from `JPanel`) class for showing the moving ants. By default, Swing components like `JPanel` are *double buffered*. Double buffering allows dynamic displays to update smoothly without an apparent flicker of the screen. This is done by drawing all graphics to an invisible offscreen buffer before transferring the completed buffer rapidly to the panel. Unfortunately for us, AWT components like the `Applet` do not use double buffering. As a result, we are forced to implement our own by using the `Image` class to hold an offscreen buffer.

Communicating with an Applet

We can add one further set of enhancements to our applet by getting the simulator to communicate with the web browser embedding it. It is considered good practice for all applets to include a method `getAppletInfo()` that returns a simple message describing the purpose of the applet. What the browser chooses to do with this message varies between browser manufacturers, but there is little cost in efficiency or resources in adding this method to every applet you create.

```
/** Provides a description of this applet for the browser.
  * @return Message describing the applet.
  */
public String getAppletInfo()
{
    return "Ants in the Garden - A Simulation by Jo Wood, 2001";
}
```

We can also get the applet to communicate with the browser while it is running using the method `showStatus()`. This allows any arbitrary message to be reported back to the browser, which will usually display this information somewhere in the browser window. We can get the `Garden` class to report the speed of the simulation by adding a method `displayMessage()` to the `GraphicsListener` interface. This can then be intercepted by the applet and reported to the browser.

```
/** Displays the given message in the applet. Some browsers may
  * choose to ignore this request.
  * @param message Message to display.
  */
public void displayMessage(String message)
{
    showStatus(message);
}
```

To use this method, the Garden class simply makes a call to `graphicsListener.displayMessage()`. Sample output from this now fully Java1.1 compliant applet is shown in Figure 10.4.

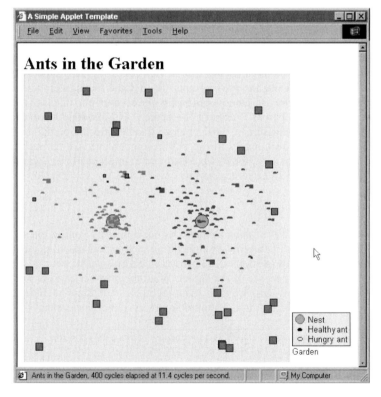

Figure 10.4 Sample output from ant simulation applet showing applet status at bottom of window.

10.4.2 Creating a Java 2 Spatial Model Display Applet

The Spatial Model display application developed in the previous chapter makes considerable use of the Java 2 API, so it is impractical to re-engineer it as a Java 1.1 applet. An alternative approach is to force the browser to use a Java 2 Virtual Machine by using the Java plug-in rather than its in-built Virtual Machine. The simplest way of doing this is to run a script in the web page hosting the applet to check for any installed plug-in and download a new one if necessary. We need not concern ourselves how such a script works, so long as we have access to one. Thankfully, Sun provide just such a script that can be called remotely from a web page. The listing below shows how a Java 2 applet may be called in this way.

```
<!DOCTYPE HTML PUBLIC "-//W3C//DTD HTML 4.0 Transitional//EN">
<html>
 <head>
  <title>Spatial Model Display Applet</title>

  <!--- Performs an automatic check that the latest Java Plug-in
        is installed. --->
  <script src="http://java.sun.com/browser/autodownload.js">
  </script>

 </head>
<body bgcolor="white">

<h1>Spatial Model Display using a Java 2 Applet</h1>

<applet code="SpatialApplet.class" codebase="packages"
        width="30" height="10">
 No Java 2 support for this applet.
</applet>

</body>
</html>
```

Calling an applet in this way should allow us to use the entire Java 2 API in our own classes. The only restrictions being the limitation on file input/output imposed by the applet *security manager*. These restrictions prevent any of the file handling routines developed in the previous chapter from being used. If any restricted file activity is attempted, the security manager will immediately halt execution of the applet in send an error message to the Java console. This is a rather 'ugly' way for our program to behave, so it is better to modify our code to prevent file I/O from being attempted in the first place.

Rather than write two versions of our classes, one for applets without file handling, and one for applications with file handling, we can define a boolean object variable `isApplet`, which determines whether or not to use any file I/O functionality. The modified sections of `ModelDisplay` are shown below.

```java
package jwo.jpss.spatial.gui;   // Part of the spatial GUI package.

// ****************************************************************
/** Creates a simple window for displaying spatial models.
  * @author   Jo Wood.
  * @version  1.3, 9th October, 2001
  */
// ****************************************************************

public class ModelDisplay extends JFrame implements ActionListener
{
    // ----------              Object Variables -------------------

    private boolean isApplet;

    // ---------------------- Constructor ----------------------
    /** Creates a top-level application window which displays the
      * given spatial object.
      * @param spObject Spatial object to display.
      */
    public ModelDisplay(SpatialObject spObject)
    {
        this(spObject,false);
    }

    /** Creates a top-level window displaying given spatial object.
      * @param spObject Spatial object to display.
      * @param isApplet Indicates whether running as an Applet.
      */
    public ModelDisplay(SpatialObject spObject, boolean isApplet)
    {
        // General initialisation here.

        this.isApplet = isApplet;

        // Perform non applet initialisation.
        if (!isApplet)
        {
            // Menu creation here.

            // Initialse the file chooser.
            fileChooser = new JFileChooser();
            fileChooser.setFileFilter(new SpatialFileFilter());
        }
    }

    // ---------------------- Methods ------------------------

    /** Attempts to open given file and update the spatial object
      * being displayed.
      * @param inFile Spatial file to open.
      * @return True if opened successfully.
      */
    public boolean openFile(File inFile)
    {
        // Don't allow applets to open files.
        if (isApplet)
            return false;
```

```java
        // Open file code here.
    }

    /** Attempts to save the current spatial object.
      * @param outFile Spatial file to save.
      * @return True if saved successfully.
      */
    public boolean saveFile(File outFile)
    {
        // Don't allow applets to save files.
        if (isApplet)
            return false;

        // Save file code here.
    }
    // ----------------- Implemented Methods --------------------

    /** Responds to a selection of menu item.
      * @param event Menu selection event.
      */
    public void actionPerformed(ActionEvent event)
    {
        // Don't allow applets access to menus.
        if (isApplet)
            return;

        // Menu handling code here.
    }
    // ------------------- Private Methods ---------------------

    /** Asks the user if they really want to quit, then closes.
      */
    private void closeDown()
    {
        int response = JOptionPane.showConfirmDialog(this,
                            "Are you sure you want to quit?");
        if (response == JOptionPane.YES_OPTION)
        {
            if (isApplet)
                dispose();           // Close window.
            else
                System.exit(0);      // Exit program.
        }
    }
}
```

After removing all file import functionality from the applet-friendly version of our class, we are faced with the problem of getting useful data (raster maps or vector maps) into our program. We have several choices here. We could use the getImage() method of the Applet class to retrieve the data used in a raster map. This could be supplemented with metadata (title, author, notes, etc.) extracted from applet parameters. This was the approach taken in Section 10.2. Alternatively, we could make the applet talk to a Java application sitting on the same web server. This application could handle all the file input/output and send a serialized version of a `SpatialObject` to the applet. This procedure requires the use of *sockets*, a technique we will not cover in this book.

The simplest solution is probably to create the `SpatialObject` programmatically within the `Applet` class. The listing below shows an example of this, along with typical applet output in Figure 10.5.

```java
import jwo.jpss.spatial.*;        // For spatial classes.
import jwo.jpss.spatial.gui.*;    // For ModelDisplay class.
import javax.swing.*;             // For Swing GUI.
// *****************************************************************
/** Starts the GIS display application..
  * @author    Jo Wood.
  * @version   1.4, 11th October, 2001
  */
// *****************************************************************

public class SpatialApplet extends JApplet
{
    // -------------------- Starter methods --------------------

    /** Creates a top-level window and simple raster and displays
      * it on its own panel.
      */
    public void init()
    {
        // Create a simple vector map to display.
        float xCoords[] = {2151645,2150994,2149195,2149142,2149541,
                           2149472,2149770,2150109,2150374,2150918,
                           2151645,2152331,2153326,2154056,2154372,
                           2154894,2156566,2157099,2157612,2158631,
                           2159228,2159827,2160106,2160144,2160131,
                           2160105,2160094,2160010,2159827,2158902,
                           2158089,2157512,2157099,2156279,2155376,
                           2154372,2153277,2152574,2151645};

        float yCoords[] =  223376,223528,224100,224296,225948,
                           226495,227244,227844,228890,229529,
                           230404,230683,231286,231564,231703,
                           231744,231754,231804,231735,231286,
                           230760,230058,229136,228890,228623,
                           226740,226495,226334,226032,224912,
                           224100,223737,223516,223379,223218,
                           223056,223138,223284,223376};

        float xLine[] = {2150000, 2153000, 2158000, 2162000};
        float yLine[] = {233800,  228000,  225000,  224000};

        Header vectHead = new Header("A simple vector map",
                    "Jo Wood",
                    "Copyright October, 2001",
                    "Simple vector map with one area and line");
```

```
            GISVector gisVect = new GISVector(xCoords,yCoords,
                                              SpatialModel.AREA,10);
            VectorMap vectMap = new VectorMap(gisVect,vectHead);
            vectMap.add(new GISVector(xLine,yLine,
                                              SpatialModel.LINE,20));
            new ModelDisplay(vectMap,true);
    }
}
```

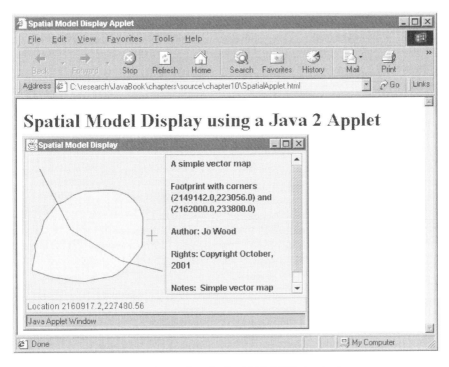

Figure 10.5 Output from the Spatial Model Display Applet.

10.5 SUMMARY

In this chapter, we have build upon the ideas and techniques developed throughout this book in order to allow our Java programs to make use of internet technologies. The way this is most commonly achieved is to use applets allowing Java programs to be embedded in web browsers. Applets have the advantage of allowing others without the Java SDK installed on their system, to view your Java programs with minimal effort.

Producing Java applets involves producing an HTML page that creates a link between the web browser and your Java classes. This link can include 'parameters' effectively allowing messages to be passed from HTML to Java methods.

Security restrictions and browser compatibility problems can limit the functionality of Java applets, especially with respect to file handling. It may also sometimes be necessary to restrict classes to the Java 1.1 API in order to maximise compatibility with web browsers.

For more sophisticated transfer of data over the internet, XML provides a platform-independent mechanism for marking up information. Java can interpret and create XML files using the SAX and DOM APIs. The DOM allows a tree of XML nodes to be created and traversed. We saw how a simple recursive search method can navigate through this tree structure in order to extract the information associated with various XML elements.

By the end of this chapter and associated exercises on the web, you should be able to

- create a simple applet embedded in a web page including the passing of parameters from HTML to your applet;
- convert an existing application (with a `main()` method) into an applet (with an `init()` method;
- be aware of the security restrictions placed on applets;
- identify when it is more appropriate to use an applet and when it would be better to use an application to achieve a particular task in Java;
- understand what an XML file is and why it might be useful in representing information; and
- create a Java program to extract an element from an XML file using the DOM.

References and Further Reading

You may find the following references and web pages useful in continuing to explore the potential of Java in the spatial sciences. Most of the references are introductory and provide good starting points for more in-depth exploration. They are listed in approximate order in which they become relevant in the chapters of this book. Further references may be found at www.soi.city.ac.uk/jpss.

CHAPTER 1 – INTRODUCTION

www.soi.city.ac.uk/jpss The website supporting this book. The site includes all source code shown in this book as well as exercises, quiz questions and further Java-related resource.

Hoyt, E. (1996) *The Earth Dwellers: Adventures in the Land of Ants*, New York: Touchstone. A very readable entomological account of ant behaviour. Its ant's perspective is particularly appropriate reading for those interested in object-oriented modelling.

www.javasoft.com Sun's official website for distribution of information about Java. Includes up-to-date news on new Java releases, downloadable software (including the Java SDK) and a comprehensive set of online tutorials.

www.javaworld.com One of the longest running online 'Java journals'. Contains a large collection of Java related articles, reviews and tutorials. Updated weekly.

CHAPTER 2 – INTRODUCING CLASSES AND OBJECTS

www.uml.org A good starting point to find out more about the *Universal Modelling Language* used for representing object-oriented models. Includes a full specification of the language and links to UML creation and diagram resources.

JavaWorld (2001), *Object-oriented language basics*, www.javaworld.com/javaworld/jw-04-2001/jw-0406-java101.html A useful series of articles on object-oriented terms and principles.

CHAPTER 3 – DEVELOPING CLASSES AND OBJECTS

Walrath, K. and Campione, M. (1999) *The JFC Swing Tutorial: A Guide to Constructing GUIs*, Addison-Wesley, also online at java.sun.com/docs/books/tutorial/uiswing. A comprehensive tutorial and information source on constructing graphical user interfaces with the Java Swing library.

Sun Microsystems (2001) *Java look and feel design guidelines (2nd edition)*. java.sun.com/products/jlf/ed2/book Some guidelines on how to use and create graphical user interfaces with Swing components.

CHAPTER 4 – CONTROLLING PROGRAM MOVEMENT

Boole, G. (1847) *The Mathematical Analysis of Logic*, reprinted by St. Augustine Press (1998). The first application of algebraic methods to formal logic.

Boole, G. (1859) *An Investigation into the Laws of Thought*, reprinted by Dover Publications (1953). The conceptualisation of formal logical principles upon which most modern computing is based.

CHAPTER 5 – MAKING DECISIONS

Gosling, J., Harrison, J. and Baritz, J. (2001) *Sorting Algorithms* www.cs.ubc.ca/spider/harrison/Java/sorting-demo.html A collection of 15 different sorting algorithms demonstrated graphically using Java applets. Each includes downloadable source code. An excellent way of appreciating the comparative efficiency of different sorting processes.

www.unicode.org Provides details on the Unicode convention for encoding text in a variety of scripts.

Molenaar, M. (1998) *An Introduction to the Theory of Spatial Object Modelling*, Taylor and Francis. A comprehensive formalisation of the way in which we can model space as objects. The theory covered in this monograph provides a sound basis for representation in an object-oriented language like Java.

Egenhofer, M. and Franzosa, R. (1991) Point-set topological relations. *International Journal of Geographical Information Systems*, 5, pp.161-174. An early formalisation of the types of topologic relations that can exist between spatial objects.

CHAPTER 6 – SHARING CLASSES

java.sun.com/docs/ The Application Programming Interface (API) for the Java language can be found online at this address. If you installed the documentation with the Java SDK, you will also have an identical copy in the docs/ folder of your installation.

Sommerer, A. (2001) *The Java Tutorial: JAR Files*, java.sun.com/tutorial/jar An introductory tutorial on how to use the Java archiving tool JAR.

geotools.sourceforge.net A large open source collection of Java classes for creating and manipulating web based maps.

CHAPTER 7 – COLLECTING OBJECTS TOGETHER

java.sun.com/jdbc *The JDBC Data Access API*. This page is a good starting point for investigating the Java Database Connectivity package. Allows platform independent connectivity with a range of relational database management systems.

Schultz, E. (2001) *The Genetic Algorithms Archive* www.aic.nrl.navy.mil/galist A good starting point for exploring evolutionary programming and genetic algorithms.

References and Further Reading 303

deBerg, M., van Kreveld, M., Overmars, M. and Schwarzkopf, O. (2000) *Computational Geometry: Algorithms and Applications. 2nd Edition* London: Springer-Verlag. See also www.cs.ruu.nl/geobook. An excellent source of algorithms for handling geometric and spatial data. The accompanying web site includes software and links to other useful sites.

CHAPTER 8 – CONTROLLING DYNAMIC EVENTS

java.sun.com/products/java-media/2D Contains information on the Java 2D toolkit for high quality graphics and image manipulation. This toolkit contains a range of useful classes for handling spatial objects.

CHAPTER 9 – HANDLING STREAMS AND FILES

www.esri.com The developers of the Geographic Information System ArcGIS. Includes useful links to other GIS-related sites and data.
www.geog.uni-hannover.de/grass The pages of the freely available Geographic Information System GRASS. Includes source code (written in C) and links to Java using the Java Native Interface.

CHAPTER 10 – COMMUNICATING WITH THE WIDER WORLD

java.sun.com/products/plugin Describes how to install and use the Java plug-in for creating an up-to-date Java virtual machine in a web browser.
www.w3.org/MarkUp Probably the best place to start for all things HTML. Includes an introduction to the language, and links to useful resources.
www.w3.org A good starting point for all new 'official' web standards such as HTML, XML, XSL etc. Includes useful links to other resources and tutorials.
xml.coverpages.org *The XML Cover Pages.* A good place to start if you wish to find out more about XML. Includes introduction, up-to-date news and links to software and other resources.
java.sun.com/xml Includes documentation for the Java-XML interfacing libraries JAXP and others.
sax.sourceforge.net The official homepage for SAX – the simple API for XML allowing programs in Java and other languages to interface with XML documents.

Glossary

Abstract class Any class that has at least one abstract method. It is not possible to create an object directly out of an abstract class without first inheriting the class and implementing the abstract methods. Abstract classes are created when some of the behaviour of a class must be specified but may be implemented in a number of different ways. If all methods are abstract then the class will usually be created as an interface.

Abstract method A method that describes an aspect of a class's behaviour, but which has not been implemented. Abstract methods are created when it is known *what* a class will do, not *how* it will do it. Abstract methods are created using the keyword `abstract`.

Abstract Windowing Toolkit Usually referred to as the AWT, this is a series of Java classes for creating a Graphical User Interface. The AWT allows the programmer to create windows, buttons, sliders, etc. and manage the events generated by each of them.

Abstraction The process of designing a class by identifying the behaviour (and therefore methods) of a class, but without implementing that behaviour. Abstraction is an important part of object-oriented design as it allows the designer to separate the identification of *what* a class does from *how* it does it. This is often carried out by creating interfaces.

Accessor method A method in a class that is used to find out the value of a field contained within it. By convention, most accessor methods have the form `get`*Name* where *Name* is a name indicative of the field being queried. For this reason, accessor methods are sometimes known as get methods.

Note that it is generally much better practice to keep variables within a class private and use accessor (and mutator) methods to query (and change) them than it is to make variables public.

Adapter class An class containing implemented methods of an interface, but which do nothing. Adapter classes are a convenience for the programmer who wishes to implement only a subset of an interface's methods. Examples include `WindowAdapter` (implementing `WindowListener`) and `MouseAdapter` (implementing `MouseMotionListener`).

Affine transformation The process of changing geometric coordinates by scaling, translating, rotating, or a combination of all three. Affine transformations are particularly useful for changing between different coordinate systems (e.g. screen coordinates to georeferenced coordinates). Java can store and perform affine transformations using the `AffineTransform` class in the `java.awt.geom` package.

Algorithm A set of instructions that describe a process or operation. These instructions are specified in an unambiguous fashion such that they could be implemented in a programming language. There are several formal languages for specifying algorithms as well as less precise 'pseudo code' that often looks similar to an existing programming language (but usually more general).

API	See Application Programming Interface.
Application programming interface	Documentation associated with a software package that allows programmers to link their own code with the software. In Java this is likely to be a description of all the public classes, fields and methods associated with one or more packages. An example might be the Swing API that allows Java programmers to create their own graphical user interfaces.
Arc	In topological terms, an arc is a collection of one or more straight line segments bounded by two nodes. Arcs are typically used to represent linear features such as sections of roads bounded by junctions, or contour lines.

Arrray	A collection of variables, all of the same type and of a fixed size. Arrays in Java can be of any type from primitives (float, int etc.) to complex objects. Arrays are identified by the use of [square brackets], both to identify the array size and individual elements within the array. Array indices start from 0 rather than 1.
	
	For example,
	
	`int results[] = new int[25];`
	
	creates an array of 25 elements, while
	
	`results[10] = 99;`
	
	places the value of 99 into the 11th element of the array.

AWT	See Abstract Windowing Toolkit.
Behaviour	The part of a class or object definition that describes what the class or object does. Behaviour descriptions distinguish object-oriented models from procedural ones (e.g. those created by languages like C and Fortran) which tend only to describe something's state.
Binary	A number system based on only two digits, 1 and 0. All computer processing ultimately involves handling binary digits, although the Java compiler and interpreter allow us to translate our own instructions from Java into binary. Java allows us to manipulate binary numbers directly using the binary operators & (binary AND),	(binary OR) and ^ (binary NOT).
Bit	A binary digit able to store either a 1 or a 0. Bits are often linked together as bytes or words, typically in groups of 8, 16, 32, and 64.
Byte	A collection of 8 bits capable of storing 256 different combinations of 1s and 0s. This is the unit by which storage capacity is measured. For larger capacities, kilobytes (1024 bytes), megabytes (1024 kilobytes), gigabytes (1024 megabytes) or even terabytes (1024 gigabytes) are used.
Class	A generic description of a group of objects that will have a state and behaviour. In Java, classes are usually defined using the class keyword and are given an initial upper-case letter.
Class diagram	A graphical representation of one or more classes and their relationship to one another. Class diagrams are often used to summarise the

Glossary

organisation of a group of classes without the need to refer directly to program code.

Class diagrams will sometimes identify all public methods and variables, in which case the diagram becomes a summary of a program's interface or API.

Class library A collection of related classes for performing a related set of tasks. Consists of one or more related packages. An example of a class library is the Abstract Windowing Toolkit used for creating Graphical User Interfaces.

Class variable A variable declared within a class but outside any method with the modifier static. Class variables may be manipulated without instantiating an object out of the class. Only a single copy of a class variable is created regardless of the number of objects instantiated. The most common use for class variables is to create constants or identifiers. For example,

```
public static final float PI = 3.141592654;
```

Composition The process of storing an object made from one class as a field of another. This is the commonest form linking classes together, and is sometimes referred to as a 'has a' relationship.

Computational geometry A branch of mathematics/computer science concerned with representing and manipulating geometrical and topological features. Forms the basis for many of the algorithms and data structures used in spatial programming.

Constant A special type of field that stores an unchanging value. Typically this might be a mathematical constant such as PI, or an identifier such as RED. In Java, a constant is indicated by using the keyword modifier final, and by convention is named with upper-case letters.

Constructor A special method declared inside a class that is automatically called whenever an object is instantiated from the class. A constructor shares its name with the class in which it is defined, and is used to initialise an object.

Data structure The arrangement of data items within some identifiable unit used to represent information of some kind. Data structures can be simple, such as a single array of integers, or more complex such as a network of arcs that make up a GIS vector model.

Declarative programming A style of programming based on the construction of a collection of rules and the posing of questions based on those rules. Unlike procedural programming, the ordering of declarative statements tends to have little effect on the result of the programming process. Can be contrasted with procedural and object-oriented programming.

Delegation The process of using message passing to transfer the responsibility for performing some behaviour from one class to another. The most widely used example of delegation is probably the event listeners used by AWT and Swing components.

Dictionary A type of dynamic group used for storing a collection of objects. Each object stored in the dictionary is identified by some keyword or other identifier.

Documentation The process of describing how a program works and how it is used. Documentation is vital when developing code, especially when

	intending to share it with others. Documentation may be internal using source code comments and external such as 'help files'. Documentation is a necessary part of producing an API.
DOM	A Document Object Model. Used to represent the information stored in an XML document. This tree structure can be manipulated relatively easily in Java.
Dynamic group	A collection of objects that can grow or shrink throughout its life. Examples of dynamic groups include the Java Vector and HashMap. Can be contrasted with fixed groups such as the array.
Encapsulation	The gathering together and hiding of all the detailed workings of a class. Encapsulation is one of central processes in object-oriented modelling and tends to simplify the modelling process by hiding unnecessary detail from the programmer. In Java, parts of a class can be encapsulated by limiting the scope of a method or variable.
Escape code	A special pair of characters used for formatting text output. Common escape codes include \n to represent a new line, \t to represent a tab stop, \\ to represent a single \ and \" to represent a quotation mark.
Event	A special type of message that is sent between objects at run-time. The most common type of events are those generated when using a Graphical User Interface, such as mouse clicks on a button, or a window being resized. Most GUIs will contain some kind of event handling to deal with such occurrences.
Exception	A special type of error generated at run-time. Exceptions cannot always be avoided as they may be a result of factors outside the programmer's control, such as disk failure or lack of memory. Exceptions can be handled in Java using the keywords try and catch.
Extension	A type of inheritance whereby a new subclass is used to add to the functionality of an existing class. See also generalisation.
Field	Part of a class or object that can be used for storing an element of its state. Will be either a variable or constant.
Floating point	A type of Java primitive used for storing numbers containing a decimal point (identified by the word float in Java). A number can be identified as being floating point by appending an 'f' to it.

For example,

```
float temperature = 25.6f;
``` |
Generalisation	A type of inheritance where common functionality from two or more classes are extracted and placed in their own superclass. See also, extension. Should not be confused with the separate process of spatial generalisation where the important characteristics of a representation are extracted and emphasised.
Get method	See accessor method.
GML	Geography Markup Language. An implementation of XML used for storing and representing spatial data. Its development has been coordinated by the OpenGIS Consortium or OGC.

Glossary

Graphical user interface	A special case of user interface (UI) whereby interaction is facilitated through the use of graphics. The most widely used form of GUI uses the windows paradigm of windows, menus, buttons etc. controlled using a mouse, pointer and keyboard. The Java class library the Abstract Windowing Toolkit (AWT) allows such GUIs to be developed.
GUI	Usually, pronounced as '*gooey*' – see Graphical User Interface.
Handle	See reference.
Hash map	A type of dynamic collection containing a set of *keys* and *values*. Each value in the collection can be any Java object and is uniquely referenced by a key that can also be any object. Similar to the Java 1.1 class – the `Hashtable`.
Implementation	The process of providing the instructions that make up the abstract methods in an interface. This allows several classes to use the same interface, but for each to implement it in its own way. The relationship between a class and the interface it implements is sometimes described as an 'acts as a' relationship and uses the Java keyword `implements`.
Inheritance	The process by which one class borrows the state and behaviour descriptions of another class and adds them to its own state and behaviour description. Inheritance is one of three ways in which classes may be linked, and is sometimes described as an 'is a' relationship. In Java, one class can inherit another by using the `extends` keyword.
Instance variable	See object variable.
Instantiate	To create an object based on an existing class definition. This is accomplished using the Java keyword `new`.
Integer	A whole number. In Java, variables declared to hold integers can use the keyword `int`. Such variables can store integers using 32 bits allowing numbers between –2147483648 and 2147483647 to be stored. Integers can also be stored and manipulated as objects using the `Integer` class.

Intercapping	The convention whereby variables are named in lower-case letters with no spaces. The first letter of any concatenated words are given an upper-case letter. For example, the following are all examples of intercapped variable names:
	
	`JavaProgram`
	
	`businessToBusiness`
	
	`infoSystems`

Interface	The result of the abstraction process whereby a class is described by what it does rather than how it does it. Good object-oriented programming will usually involve defining a class' interface before any coding is carried out. An interface must be implemented before it can be used as an object. In Java, an interface will consist of a class name and a list of all public methods within the class. The Java keyword `interface` allows the programmer to declare a class and its methods without specifying the details of how the methods do their job. Declaring interfaces can help in creating reusable classes.

Iteration	The process of repeatedly performing a task. Structures such as the `for` loop and `do-while` loop perform iteration. Not to be confused with *recursion*.
Iterator	Something that repeatedly performs a specific task, usually counting through a collection of objects. If the collection is a dynamic group such as a `Vector`, the size of the group is often not known at compile-time. Therefore, an iterator can be used to count through the elements of a collection testing to see whether there are any more elements left to count. Java has a special `Iterator` interface that allows most dynamic groups to be counted using the method `hasNext()`.
javadoc	The program tool supplied with Sun Microsystems' Java implementation that allows program documentation to be generated (semi) automatically from inline comments. Any Java source code that is enclosed by `/**` and `*/` can be used to create documentation with `javadoc`. Can be used to generate documentation of a class' public interface. When a collection of classes are documented in this way, is sometimes known as an API or Application Programming Interface.
Keyword	A word that is part of the Java language and may not be used by the programmer in any other context (such as variable names). Examples of keywords include `boolean`, `if` and `switch`.
Linked list	A data structure for representing a dynamic collection whereby each element in the collection contains a reference to one or more other elements in the same collection. This allows an iterator to 'traverse' the elements of a collection quickly and efficiently.
Message	Data passed either to or from an object. Messages sent to an object will usually use the parameters of a method, while data sent from an object will usually use the return value from a method.
Method	A part of a class that represents a particular aspect of its behaviour. All methods have a name which should be indicative of the behaviour they represent, followed by brackets (). Optionally, methods can also accept incoming messages in the form of one or more parameters, and one outgoing message indicated by the `return` statement.
Multiple inheritance	The process of a class inheriting more than one class simultaneously. Unlike C++, multiple inheritance is not allowed in Java (the `extends` keyword can only be followed by a single class name). This is not usually a problem as Java uses a 'single rooted hierarchy' of classes – that is all classes are inherited from the class `Object`. Multiple interfaces can be implemented.
Mutator method	A method in a class that is used to change the value of a field contained within it. By convention, most mutator methods have the form `setName` where `Name` is a name indicative of the field being changed. For this reason, mutator methods are sometimes known as set methods. Note that it is generally much better practice to keep variables within a class private and use mutator (and accessor) methods to change (and query) them than it is to make variables public.

Glossary   311

Node   A point in a GIS vector model that identifies either the end of an arc, or the junction of one or more arcs. The term is also used to describe elements in a variety of data structures such as trees and linked lists.

Null   An indication that a class or variable is empty, or that a method has returned an empty message. Classes that have not been instantiated as objects will have the value 'null' as will variables that have been declared but not initialised. Java can be used to manipulate null conditions with the keyword null.

Object   A clearly defined model of something that has a state and behaviour. An object is an instantiation of a more generic class. In Java, objects are usually created using the keyword new and are given an initial lower-case letter.

Object variable   Sometimes referred to as an instance variable. An object variable is one that is declared inside a class but outside any method definition. This makes its scope available to all methods within a class. Object variables are often prefixed with one of the modifiers, private, protected or public.

Override   To redefine the body of an existing method or contents of an existing variable from an inherited class.

Package   A collection of related classes that form a coherent group of some kind. When creating a class, it can be placed inside a package by using the Java keyword package.

Polygon   A 2D shape with at least three sides. Is often used in spatial sciences to represent the boundaries of areas, and by implication, their contents.

Polymorphism   The handling of objects of 'many shapes'. A method can receive an incoming message of a range of object types, as long as all objects inherit the same superclass. For example, if we define three classes as follows

```
public class Animal() {}
public class Cat() extends Animal{}
public class Dog() extends Animal {}
```

We can create a method in another class such as

```
public void makeNoise(Animal animal) {}
```

We could pass either the Cat or Dog to this method as they both inherit Animal.

Primitive   A fundamental unit of numerical storage in Java. Primitives include the types, boolean, char, short, int, long, float and double.

private   A Java keyword indicating that a given variable, method or class cannot be used outside the scope in which it was defined. In practice, this means that if something is declared as private it cannot be seen or changed from outside the class in which it occurs. This is one way in which Java can be made to encapsulate the details of state and behaviour within a class. See also, protected and public.

Procedural programming A style of computer programming where an ordered series of instructions are given to the computer. Procedural programs can usually be read 'top-down', where the order of instructions determines the order

	in which they are executed by the computer. Can be contrasted with declarative and object-oriented programming.
protected	A Java keyword indicating that a given variable, or method can only be used by the class in which it was declared, or by any of its subclasses. In practice, this means that if something is declared as protected the only classes that can use it are the one in which it was declared or any subclasses of it. This is one way in which Java can be made to encapsulate the details of state and behaviour within a class. See also, private and public.
Pseudo code	A semi-formal description of the code required to perform some behaviour or describe the structure of a class. Can be used as an intermediate stage between a class or algorithm's design and its implementation in a programming language.
public	A Java keyword indicating that a class, method or variable is accessible from outside the class in which it has been defined. As a very general rule, class and object variables should not be public unless you have good reason to allow direct manipulation from other objects. It is better to allow the manipulation of such variables using public accessor methods and mutator methods.
Queue	A data structure used for storing an ordered collection of data items. As with a real queue of people, items are added and removed from a queue on a 'first in, first out' basis. Can be thought of as adding items to the *rear* of the queue and removing them from the *front*. See also the stack.
Recursion	The process of a method or class referring to itself when performing a particular task. For example,

```
public boolean search(Node treeNode, String txt)
{
    String name = treeNode.getName();
    if (name == null)
        return false;
    if (name.equals(txt))
        return true;
    else
        return search(treeNode.getNext(),txt);
}
```

	Recursive programming is a very powerful and efficient way of performing certain tasks, but can be difficult to debug if programmed incorrectly.
Reference	A pointer to an object, sometime referred to as a handle. When an object is passed as a message from one method to another, the whole object is not passed, but rather a reference to it, saving time and memory. Unlike languages like C and C++, references or pointers cannot be manipulated explicitly.
Reusability	The degree to which the same program code may be used in a variety of contexts. Efficient and well-designed code often does more than solve a specific problem, and is written and documented so that it may be reorganised to solve new programming tasks. When designing Java classes, possible reusability should be considered from the outset. One of the most important aspects of reusability is good clear class

Glossary

documentation that describes what the class does.

Scope	The degree to which a class' methods or fields can be called or examined by other parts of a program. Two factors determine the scope of a method or field, where within a class it is declared, and the use of the keyword modifiers, public, private and protected.
Segment	In the GIS vector model, a straight line defined by two sets of coordinates, one at each end. Segments are usually combined to represent more complex lines or arcs.
Set method	See mutator method.
Signature	The type of incoming messages or parameters that are passed to a method. A class can define several methods of the same name providing they all have unique signatures. In other words, each must have a different number or type of incoming messages.
Sort	The process of arranging items in a collection into some defined order. There are many ways of sorting, the suitability of which will depend on the nature of the collection. Common sorting techniques include the bubble sort and quick sort.
Stack	A data structure used for storing an ordered collection of data items. As with a vertical stack of bricks, items are added and removed from a stack on a 'last in, first out' basis. Items can only ever be added or removed from the *top* of the stack. See also the queue.
State	The part of a class or object definition that describes what information is contained within the class or object. Can be contrasted with a class or object's behaviour.
Static variable	Sometimes called a 'class variable', this is a field or method whose contents are associated with an entire class rather than a specific object. Static variables are often used to store constants that will not change between instantiations. In Java, static variables and methods are indicated with the keyword `static`.
String	A collection of characters which together are used to represent text. In Java, strings can be stored and manipulated using the `String` class, which includes several useful methods for handling text, such as string comparison, concatenation and lower to uppercase conversion.
Subclass	A class that has inherited the state and behaviour of another (which is known as a superclass). Subclasses can be created by using the Java keyword `extends`.
Superclass	The 'parent' of a subclass. All public or protected methods and variables of a superclass are available to any subclasses that inherit it.
Thread	A single sequential flow of program instructions. The concept of threads becomes more useful when multiple threads are used in the same program. This allows instructions to be processed in parallel. Commonly, multiple threads are used to speed up the responsiveness of GUIs, and to respond to dynamic events. In Java, threads are created explicitly either by inheriting the `Thread` class, or by implementing the `Runnable` interface.
Tree	A type of data structure that links elements or nodes together without any closed loops. Often used for representing hierarchical structures,

and can be navigated efficiently using recursion. The DOM structure for representing XML is an example of tree.

Unicode A coding system for representing text characters in a variety of international scripts (Latin, Cyrillic, Hebrew, Arabic, Hiragana etc.) Java uses Unicode to store text in its `String` class and `char` primitive.

User interface The means by which communication between the user and computer is enabled. This will usually involve some form of input from the user, and output from the computer. The nature of the user interface will depend on purpose of the application, user requirements, and hardware constraints. It may involve simple text-based user interaction (such as on a mobile phone) or a more sophisticated graphical user interface (GUI). This term should not be confused with the object-oriented concept of the interface.

Variable A portion of the computer's memory that can be used for storing, changing and retrieving an element of data. In object-oriented programming, variables are often described as fields. They can range from the simplest Boolean values, to complex objects containing many items of data.

Vector In the context of Java programming, a `Vector` is a class in the `java.util` package for storing lists of objects. A vector list can be added to or reduced dynamically at run time.

Spatial programmers may also recognise a (GIS) vector as a representation of a spatial boundary by one or more sets of coordinates. Just to confuse matters further, Java `Vectors` can conveniently be used for representing GIS vectors.

Word A collection of bytes used to represent larger units of data. Words are typically 2, 4 or 8 bytes long. Different computer systems may use different length words as their basic unit of numeric processing (so-called 32 bit or 64 bit machines for example) although Java standardises its storage units across platforms.

XML eXtensible Markup Language. A 'meta-language' for creating new markup languages such as HTML. The use of XML has generated considerable interest in recent years as it allows the relatively easy representation and transmission of information over the internet.

Index

9-intersection model, 127

`abstract` (Java keyword), 48
abstract data type, 50
abstract methods, 48
abstract windowing toolkit, 55, 137
    component hierarchy, 57
abstraction, 6
accessor methods, 52
`ActionListener` (Java interface), 196
`actionPerformed()` (Java method), 196
adapter class, 199
adapter methods, 217
ADT. *See* abstract data type
affine transformation, 219
`AffineTransform` (Java class), 219, 276
aggregation, 21
algorithm, 75, 106
anti-aliasing, 224
ants in the garden, 2, 7, 62, 117, 142, 173, 289
API. *See* application programming interface
applets, 10, 265, 289
    security, 271
    trusted, 272
`appletviewer` (Java tool), 273
application programming interface, 135, 272
    documentation, 137
ArcGrid, 251
archiving, 141
ArcInfo, 251
arithmetic operators, 73, 74
array, 84, 155
    declaration, 85
    element, 85
    index, 85
    initialisation, 85
    multidimensional, 86
ASCII, 114
assembly language, 6
assignment operation, 75
attributes, 280
AWT. *See* abstract windowing toolkit

behaviour, 8, 19, 31
binary numbers, 4
bit, 233
bitmap, 87
BlueJ (integrated development environment), 14
Boole, George, 76
`boolean` (Java keyword), 76
Boolean expressions, 73, 76
border layout, 92
`BorderFactory` (Java class), 101
boundaries, 155
bounding rectangle, 158
braces, 16, 33
`break` (Java keyword), 112
bubble sort, 106
buffering, 235
bytecode, 14, 142, 236

C (programming language), 6, 12
C++, 7, 10, 12
card layout, 93
Cartesian referencing, 89
`catch` (Java keyword), 236
class, 16, 17
    and objects, 22
    behaviour, 19, 31
    child, 45
    definition, 19
    design, 19
    diagram, 20
    state, 19, 23
    variable, 122, 133
classpath, 142
closable window, 60
`Collection` (Java interface), 178
collection framework, 159

collections, 159
  traversal, 162
colour, 97
comments, 16, 34
compile-time control, 193
compile-time errors, 236
compiling Java, 14
composition, 21, 36, 43, 65
concurrent programming, 203
conditional expressions, 103
conditional operator, 115
constant, 122, 185
constructor, 32, 38, 48
container, 56, 94
content pane, 62, 89

data store, 6
database connectivity, 168
declarative programming, 6, 73
decrement operator, 75
delegation, 142, 149
delegation event model, 194
delimiter, 248
Delphi, 7
desktop, 11
document object model, 282
documentation, 34, 135, 137
DOM. *See* document object model
double buffering, 293
`do-while` loop, 81
dry running, 110
DTD, 280
dynamic collections, 159
dynamical systems, 74

editor, 13
empty constructor, 163
encapsulation, 9, 50, 131
entity-relationship diagram, 23
equality operator, 75
error stream, 114
errors
  compile-time, 236
  runtime, 236, 290
escape codes, 31
event handler, 194
event handling, 149, 193, 216
event listening, 149, 194

evolution, 182
evolutionary programming, 182
exceptions, 236
`extends` (Java keyword), 45
extensible markup language, 10, 279
extension, 44

feedback, 74
field, 23
file chooser, 245
file embargo, 272
file filter, 246, 262
file formats, 251
files
  binary, 243
  closing, 243
  opening, 241
  reading, 240
  serialization, 243, 249
  text, 243
  writing, 240
`final` (Java keyword), 122
`finally` (Java keyword), 237
floating point numbers, 27, 78
flow layout, 91
`Footprint` (Java class), 117, 123, 156, 210
`for` loop, 79
Forte for Java (IDE), 14
Fortran, 6

gazetteer, 168
generalisation (inheritance), 44, 121
genes, 182
georeferencing, 87
GIF, 246, 274
`GISVector` (Java class), 156, 163, 210
graphical classes, 55
graphical components, 194
graphical techniques, 2
graphical user interface, 62, 90, 93, 206, 227, 234, 243, 260
`Graphics` (Java class), 96, 98
graphics context, 96
`Graphics2D` (Java class), 223
GRASS (GIS), 251
grey-scale, 97
grid layout, 92
gridbag layout, 93

Index 317

has a relationship, 21, 43, 65
`HashMap` (Java class), 168, 169
header, 16, 34
heavyweight components, 58
helper classes, 252
helper methods, 133
high-level languages, 5
HTML, 135, 268

IDE. *See* integrated development environment
identifier, 123
`if` statement, 103
`if-else` construction, 109
`Image` (Java class), 97, 274, 293
images, 269
implementation, 64
`implements` (Java keyword), 49
`import` (Java keyword), 62, 138
importing packages, 62
increment operator, 75
indentation, 33
inheritance, 10, 43, 64, 187
    constructors, 48
`init()` (Java method), 268
input stream, 233
input, process and output, 25, 233
installing Java, 13
instance, 22
instance variables, 122, 132
`instanceof` (Java keyword), 181
instructions, 6
integrated development environment, 13, 14
intercapping, 26, 33
interface (object-oriented), 49, 64, 121, 147
International Symposium on Spatial Data Handling, 11
internet, 12, 265
interoperability, 251
is a relationship, 43, 64
iterator, 162, 176

`jar` (Java tool), 141
Java
    API, 2
    console, 290
    database connectivity, 168
    interpreter, 142
    plug-in, 273, 295
    runtime environment, 13
`java` (Java tool), 15
Java 1.0, 12
Java 1.1, 12, 161, 272, 276
Java 2, 12, 161, 272
Java2D, 221, 258
`javac` (Java tool), 15, 29
`javadoc` (Java tool), 34, 135, 272
JavaScript, 11
JavaWorld, 14
JAXP, 281
`JButton` (Java class), 95, 194
JDBC. *See* Java database connectivity
`JFileChooser` (Java class), 245, 262
`JFrame` (Java class), 62
`JLabel` (Java class), 62, 95
`JMenu` (Java class), 198
`JOptionPane` (Java class), 199
`JPanel` (Java class), 95
JPEG, 246, 274
`JProgressBar` (Java class), 207
JRE. *See* Java runtime environment
`JScrollPane` (Java class), 198
`JTextArea` (Java class), 198
`JTextField` (Java class), 234

key, 168
`KeyboardInput` (Java class), 80, 234, 238

layout manager, 91, 98
layout of code, 33, 34
lightweight components, 58
list, 86
`List` (Java class), 161, 163
list sorting, 104
listeners, 147, 149, 217
local variables, 52, 104, 131
logical operators, 73, 76
look and feel, 56
loop counter, 79
loops, 78
low-level languages, 5

machine code, 4
`main()` (Java method), 16, 31
`Map` (Java class), 161
map applet, 274
matrix, 86, 89

menu, 198
MER. *See* minimum enclosing rectangle
message
   parameters, 52
   passing, 50
   returning, 52
   sending, 51
   type, 52
methods, 17, 31
   abstract, 48
   accessor, 52
   helper, 133
   invoking, 35
   mutator, 52
   overriding, 96, 189
   private, 51
   protected, 51
   signature, 52
   variables, 132
minimum enclosing rectangle, 165
modulus operator, 74
mouse listening, 217
multidimensional arrays, 86
multiple inheritance, 203
mutation, 184
mutator methods, 52
mutually exclusive, 109

namespaces, 140
natural language, 4
nested if construction, 109
nesting loops, 83
Netscape, 11, 12, 266, 272
new (Java keyword), 35
node (DOM), 287

object, 22, 35
   instantiating, 35
   serialization, 249, 256
   variable, 122, 132
Objective-C, 7
object-oriented modelling, 7
operators, 73
origin, 87
output stream, 233
overriding methods, 47

pack() (Java method), 62

package (Java keyword), 139
packages, 137
   archiving, 141
   creation, 139
   importing, 62
   namespaces, 140
package-wide scope, 133
paint() (Java method), 145, 276
paintComponent() (Java method), 96, 148, 223
paintImmediately() (Java method), 148
Panel (Java class), 95
parallel processing, 206
parameters, 52, 88, 271, 275
parent class, 45
park model, 19, 44
platform independence, 10
Point2D (Java class), 221
point-in-polygon test, 226
precedence, 77
prime numbers, 201
primitives, 27, 28, 114
private methods and variables, 45
procedural programming, 6, 11, 25, 73
program
   extendibility, 9
   simplicity, 8
programming language, 4
programming styles, 5
progress bar, 207
Prolog, 7
protected (Java keyword), 46
protected variables and methods, 46
public methods and variables, 45
Pythagoras' theorem, 24

queen ant, 186
quiz, 204

raster, 87
RasterMap (Java class), 93, 210, 251
reading files, 240
real numbers, 27, 78
Rectangle (Java class), 145
relational databases, 23
rendering hints, 223
resolution, 87

Index

return (Java keyword), 52
reusability, 63
rotation, 220
rounding, 78, 116, 117
Runnable (Java interface), 201
runtime control, 193
runtime errors, 236, 290

SAX, 281
scaling, 219
schema (XML), 280
scope, 131
screen output, 30
scroll pane, 198
SDK. *See* software development kit
security manager, 295
semantic markup, 279
Serializable (Java interface), 249, 256
serialization, 243, 249
Set (Java class), 161
shorthand, 75
Simula, 7
skeleton class, 35
Smalltalk, 7
socket, 298
software development kit, 13
sorting, 104
   algorithms, 106
spatial
   comparison, 158
   data modelling, 2
   model, 210, 295
   modelling, 155
   transformation, 89, 143, 218
SpatialModel (Java interface), 120, 210
SpatialObject (Java class), 64, 117, 156, 165, 210
SpatialPanel (Java class), 215, 219
SQL. *See* structured query language
standard input, 234
state, 19
static (Java keyword), 40, 122, 133
status bar, 215
stream, 233
stream reader, 235
String (Java class), 108, 114
string tokenizer, 248
strings, 247

stroke, 223
structured query language, 168
subclass, 44, 45
Sun Microsystems, 12
super (Java keyword), 48, 62
superclass, 45
Swing, 55
   component hierarchy, 59, 60
switch (Java statement), 111
syntax highlighting, 13

tabbed pane, 93
tags, 268, 280
TextListener (Java interface), 234
this (Java keyword), 88
threads, 200
   creation, 201
Timer (Java class), 204
topologic relationship, 124, 127
toString() (Java method), 120
transformation, 96
transient (Java keyword), 258
translation, 219
TreeMap (Java class), 159, 161, 168
try (Java keyword), 236
Turing machine, 5
Turing, Alan, 5
typecasting, 28, 160

UML. *See* universal modelling language
unicode, 27, 114
universal modelling language, 21

variables, 23
   abbreviated names, 26
   declaration, 27, 237
   global, 133
   initialisation, 27, 40, 237
   local, 52, 104, 131
   naming, 25, 33
   private, 50, 133
   protected, 133
   public, 133
   scope, 131
   types of, 26
Vector (Java class), 159, 161, 175
vector map, 221
vector modelling, 155, 163

`VectorMap` (Java class), 210, 251
vectors
   compatibility with Java 1.1, 161
virtual machine, 13, 142, 272
void messages, 52

web pages, 268
while-do loop, 82
wildcard, 138
`Window` (Java class) 62
windowing environment, 193
`WindowListener` (Java interface), 199
windows, 11

'write once, run anywhere', 10
writing files, 240

Xerces parser, 281
XML. *See* extensible markup language
   parser, 281
   stylesheet language transformation, 10, 280
XSL, 280
XSLT. *See* XML stylesheet language transformation

ZIP compression, 141